WHITE GOATS, WHITE LIES

WHITE GOATS WHITE LIES

THE MISUSE OF SCIENCE IN OLYMPIC NATIONAL PARK

R. LEE LYMAN

THE UNIVERSITY OF UTAH PRESS

SALT LAKE CITY

∞ Printed on acid-free paper

Lyman, R. Lee.

White goats, white lies : the abuse of science in Olympic National Park / R. Lee Lyman.

p. cm.

Includes bibliographical references (p.) and index.

ISBN 0-87480-555-4 (alk. paper)

1. Rocky Mountain goat—Washington (State)—Olympic National Park.

2. Wildlife management—Washington (State)—Olympic National Park.

I. Title.

QL737.U53L93 1998

639.97'9648—dc21 97-46727

CONTENTS

PREFACE

In October 1993 Roger and Cathy Sue Anunsen, Keith and Antje Gunnar, and I went out to dinner in Seattle. I had just finished presenting the results of a research project in which I documented the inadequacy of the fossil and archaeological evidence for concluding that mountain goats were never present in the Olympic Mountains prior to 1925. While the Anunsens and I had met briefly before, this was our first opportunity to compare our perceptions of how Olympic National Park was managing the mountain goats found on park lands. As we shared stories, our disbelief grew. Why was the public apparently being misled and being denied access to what we believed to be public information?

The public—through its tax payments—was subsidizing the activities of researchers affiliated with Olympic National Park, part of the U.S. Department of the Interior's National Park Service. Olympic National Park personnel said that the scientific evidence was thorough and conclusive. That was not at all our perception. While the five of us disagreed on some of the implications of the scientific evidence, we clearly shared the opinion that someone had to point out the discrepancies between what the agency was saying about the scientific evidence and what that scientific evidence actually seemed to indicate.

All of us had been involved with the controversy surrounding the management of mountain goats on the Olympic Peninsula for several years. Being professional wildlife photographers, the Gunnars owed a substantial portion of their livelihood to the goats. They knew the back country of the Olympics as well as anyone, and they had discovered some evidence that seemed to contradict the findings of the park researchers.

The Anunsens were appointed to the park's advisory committee on mountain-goat management in 1987, but quickly found that "management" of the mountain goats did not seem to be the correct term: "elimination" was the operative word. They attended several park-sponsored open public hearings on the controversy, only to be told that the public could not ask certain questions and in some cases could not even comment on the controversy. The park service responded to their efforts to question this type of management of public resources with half-truths and apparent attempts to hide critical information.

My involvement in the controversy began because of an interest in the scientific evidence. In the 1970s and 1980s the park personnel believed that mountain goats had never lived in the Olympic Mountains and that the goats that were currently found there were the descendants of goats introduced in the 1920s. Thus, those goats were aliens and exotics—and National Park Service policy is that exotics are to be removed from park lands.

Given my understanding of the historic biogeography of North American mammals and my knowledge of the fossil record, I suspected that the belief that goats had never lived in the mountains was founded on negative evidence. I published a paper to that effect in 1988. The firestorm of activity that my paper helped create in the early 1990s prompted me to examine other lines of evidence that park personnel were using to justify their plans to eradicate the goats before the turn of the century. Again, I found that this evidence was not always what park researchers said it was. This volume reviews and evaluates much of the evidence.

I strongly believe that the personnel of Olympic National Park have misled the public and at times have published what I consider to be white lies in their efforts to muster public support for their management plans. Because I am not a park service employee, I am outside the circle of interdepartmental memos, political agendas, and scientific research. Therefore, my perception of the controversy is no doubt incomplete. If this volume leads to discussion of *all* relevant evidence and prompts the public and the scientific community to ask pointed questions about wildlife-management policies in Olympic National Park and on public lands in the United States in general, I will be pleased.

Writing this book would not have been possible without the help of numerous individuals. I thank Kenneth A. Ames (Department of Anthropology, Portland State University), Robert C. Dunnell (Department of Anthropology, University of Washington), Donald K. Grayson (Department of Anthropology, University of Washington), Madonna A. Moss (Department of Anthropology, University of Oregon), Michael J. O'Brien (Department of Anthropology, University of Missouri), Mark

R. Ryan (School of Natural Resources, University of Missouri), Randall F. Schalk, and Chris Szuter (University of Arizona) for encouragement and discussions on several issues; Paul Gleeson (Olympic National Park archaeologist), Lawrence C. Bliss (Department of Botany, University of Washington), and the Anunsens for references; and Olympic National Park for providing copies of various unpublished reports. The personnel of The Mazamas (Portland, Oregon) were most helpful in allowing me to sift through their extensive collection of historical documents. Jay Kirkpatrick (via the Anunsens) provided a copy of the 1992 report on contraception and sterilization.

Various drafts of this manuscript were read by Roger and Cathy Sue Anunsen, Russell W. Graham, Charles E. Kay, Paul S. Martin, Michael J. O'Brien, Chris Szuter, Frederick H. Wagner, Jeff Grathwohl, and several anonymous reviewers. Their comments and suggestions helped in many ways to make the volume better. Roger Anunsen kept me up to date on the late-breaking news as I worked on the final draft in 1997; the park had stopped answering my letters and inquiries in 1994. I especially appreciate Jeff Grathwohl's interest in this project and Frederick Wagner's encouraging words. Keith and Antje Gunnar graciously provided the slides from which the photographs were printed. Robert Church printed the photographs, and Dan Glover produced the maps and graphs. Finally, I thank Kathleen Lewis for her outstanding editorial skills.

At the time of this writing, there are reportedly still a few mountain goats in the Olympic Mountains, eating and trampling some of the park's plants and rolling around in wallows. This volume is not meant to resolve the issue of whether their continued presence is desirable or not. It is meant to provide a review and evaluation of the scientific evidence in the controversy and to show how that evidence has been presented to the public. It is written from the perspective of a player in the controversy who is outside the offices of the park and the policymakers.

1

INTRODUCTION

"The mountain goat, *Oreamnos americanus*, is a large, introduced herbivore that is changing the alpine ecosystem of Olympic National Park. Although native to the Cascade Mountains in central Washington, there is no evidence of goats in the Olympic Mountains of northwestern Washington before 1925 [when they were introduced by wildlife managers]" (Olympic National Park 1981:5). Thus begins the first of five unpublished management documents produced in the 1980s by biologists working for Olympic National Park. The biologists state that the mountain-goat population is so large that plant productivity is on the decline from grazing and trampling and areas denuded of vegetation or used as dust baths and wallows by the goats are experiencing unprecedented erosion. The goats appear to be significantly altering the alpine ecosystem that the park was designed to protect and preserve.[1] As a result, park managers have argued that the mountain goats must be eradicated (Olympic National Park 1995).

From the 1950s to the 1970s park managers used the goats as a feature to attract tourists. What is the basis for now arguing that the goats should be eliminated? Park biologists and managers believe that the goats are not native to park lands and are a serious ecological threat to the native biota, based on what is purported to be solid scientific research. Various individuals have questioned the scientific research, leading to a controversy over the ultimate fate of the goats.

[1] For an excellent and entertaining introduction to mountain-goat ecology, see Chadwick (1983). For more scientific treatments, see Rideout (1978), Rideout and Hoffmann (1975), and Wigal and Coggins (1982). For a good summary of mountain goats in Washington, see Johnson (1983).

NATIONAL PARKS, RESOURCE MANAGEMENT, AND SCIENCE

National parks in the United States were intended to protect and preserve in perpetuity outstanding examples of nature's handiwork for the benefit of the public (e.g., Sellars 1989, 1992, 1993; Stottlemyer 1981; Wagner et al. 1995:19–39; Wright 1992). But the focus was initially on scenery rather than ecological principles (see chapter 2). Policies geared toward ecosystem management came much later, beginning at a time when our understanding of ecology was poor due to the paucity of scientific research on relevant topics (Zube 1996). Over the years, policies have been modified on the basis of increased scientific knowledge, better reflecting the intent of the original laws: to provide natural areas for the enjoyment of the people.

But the pieces of landscape designated as parks cannot be isolated and ignored and still be expected to meet the intent of public policies. These areas must be managed; they must be protected, for example, from processes that would alter their natural state. As Frederic Wagner et al. (1995:10) point out, the values placed on the parks by the public are what ultimately drive management activities. Society's values establish management goals, which in turn result in legislation that becomes public policy. "There should be a close correspondence between values and goals: goals should be set to satisfy public values....the central goal of natural-resources management in the [National Park] System [is still, as it originally was,] preserving intact ecosystems" and, more recently, protecting threatened and endangered species, biodiversity, and ecological processes (Wagner et al. 1995:15–17).

"Resource-management agencies are constituted to advocate the social values that they are chartered to satisfy. In this day and age they also use science to enlighten policy decisions" (Wagner et al. 1995:113). Scientific research is merely a tool to help make wise management decisions, so that the actions that are implemented will result in attainment of management goals. "Management is the application of science.... Science does not advocate goal and policy directions" (Wagner et al. 1995:156). "If science is to serve management effectively, it must strive for objectivity, which requires distancing itself from management's value positions" (Wagner et al. 1995:113). Thus, the National Park Service (NPS) "must be committed to an open, objective research program in which investigators are free to pursue truth uninfluenced by the policy and bureaucratic predispositions of the organization" (Wagner et al. 1995:98).

This ideal situation is seldom attainable in the real world. For example, scientific research has shown the fallacy of some management decisions (e.g., Chase 1987; Hess 1993; Kay 1990), yet neither management policies nor goals have been modified. Sometimes economic considerations—and thus ultimately political implications—get in the way (e.g., Fischer 1995). Failure to attain the ideal may be the result of personal factors, such as a park superintendent who does not want to "get into trouble" as a result of scientific research (Wright 1992:191) or a park researcher who seeks to avoid the disapproval of coworkers (Mattson 1996:769). Ambiguous or internally contradictory policies have also contributed to the lack of goal attainment, and suggestions have been made as to how such policies might be clarified (e.g., Houston and Schreiner 1995; Wagner et al. 1995).

Scientific research as a tool for management of national-park lands has come under increasing scrutiny in recent years (e.g., Chase 1987; Hess 1993; Wagner et al. 1995; Wright 1992). Perhaps this is because the management decisions being made do not always produce the desired result. Whatever the reason, it is clear that, given legal mandates, the NPS "must be willing to consider [scientific] research results openly and objectively in the course of policy formulation" (Wagner et al. 1995:98). But this, too, is not always the case. In their survey of the operation of the NPS, Wagner et al. (1995:90, 97, 98) found much scientific research that remained unpublished and was not peer reviewed, research that was written for publication by public-relations personnel or other individuals with minimal or no training in the natural sciences and little knowledge of complex ecological issues, and research that was poorly planned from the outset so that the results had minimal direct bearing on the management issue at hand.

Reasons for this can be found in the constantly changing personnel among park administrators and researchers and the typically inadequate operating and research budgets (Wagner et al. 1995; Wright 1992). This volume, however, is concerned with the use of scientific research in the service of wildlife management. To quote Wagner et al. (1995:156) again, "the quality and effectiveness with which management is carried out are inevitably functions of the quality of its underlying science." If the science is not solid, then management decisions will not be well founded.

Further, if the science is not objective in the sense that it is divorced from a preconceived management goal, then management decisions may not be correct ecologically or in terms of policy goals.

One of the NPS wildlife policies concerns "exotic" plants and animals. These are defined as organisms that are found in an area as a result of human activities and are contrasted with "native" organisms, which are present as a result of natural rather than human processes. Exotic organisms are a regular source of concern among national park managers, especially when an "invasion" by such an organism is observed (e.g., Stone and Loope 1996; Wuerthner 1996). Use of these definitions to guide management activities has resulted in decisions not to transplant certain species into particular parks and to eradicate populations of some species from other parks. The key to implementation of this policy, of course, resides in determining if a particular species is native or exotic. For example, mountain goats were introduced in the 1940s and 1950s to areas north of Yellowstone National Park. As of 1991, descendants of those goats were "on the verge of colonizing the park" (Laundré 1991:8). This presents a dilemma for park managers because there is no historical evidence that goats were ever native to park lands, though they are native to nearby areas. Should they be allowed to colonize the park or not?

Similarly, in the 1950s bison (*Bison bison*) were introduced to areas of Alaska that were incorporated into Wrangell–St. Elias National Park and Preserve in 1980 (Peek et al. 1987). The fossil record suggests that bison were present in the general area as recently as 500 years ago, but the record is too poorly known to determine if bison existed there into the nineteenth century. Should the extant bison be considered native or exotic? The official NPS position is that they are exotic. The burros (*Equus asinus*) in Grand Canyon National Park and the feral horses (*Equus caballus*) in the Ozark National Scenic Riverways are easily perceived to be exotic because they are the descendants of domestic animals (Houston and Schreiner 1995; see Wagner et al. 1995:40 for other examples). But ambiguous cases also exist.

Are the elk (*Cervus elaphus*) of Mount Rainier National Park the native form or an exotic form (e.g., Bradley and Driver 1981; Schullery 1984)? And what about the mountain goats that threaten to colonize Rocky Mountain National Park (Wagner et al. 1995:41)? The species was declared to be native to Colorado in 1993, but this does not tell us whether it was native to park lands or whether goats existed in the state as a result of natural processes at some time in the past (Armstrong 1972). In such cases, only more scientific research will clarify the native or exotic status of the organisms. Such research seems, for example, to have clarified the issue of the native status of mountain goats in Colorado (Irby and Chappell 1994) and in Oregon, where the species has long been considered exotic (Matthews and Coggins 1994).

One such ambiguous case—the mountain goats of Olympic National Park in Washington state—is the subject of this book. The manner in which this particular case has been publicized by popular-science writers (e.g., Flanagan 1995; Luoma 1993; Sleeper 1990; Weingrod 1994; Wright 1996) and park biologists (e.g., Houston and Schreiner 1995) might lead one to believe that it is not ambiguous: according to these writers, the weight of the scientific evidence unequivocally indicates the goats are not native to park lands. Not only is the available evidence often weak and sometimes contradictory, however, but it has been selectively presented to the public and professional communities. In other chapters I delve into the details of the relevant scientific research. In the remainder of this chapter I summarize the history of the park, the goats, and the research that has emerged from the efforts of park biologists to develop a management plan.

A HISTORY OF OLYMPIC NATIONAL PARK AND ITS MOUNTAIN GOATS

In the summer of 1889 Judge James Wickersham hiked through a part of the Olympic Mountains (Wood 1968:79), apparently walking about 20 miles up the Skokomish River (Schultz 1993:28, 1994:276; see fig. 1.1). During July and August of 1890 Wickersham and five members of his family hiked up the Skokomish again, passing the army's Lt. Joseph O'Neil and his party, who were also exploring the Skokomish; they returned to Puget Sound, apparently via the eastward-draining Dosewallips River (Mathias 1928:37). Wickersham was so enamored with the scenery that he wrote letters on November 3 and 8, 1890, to two publishing companies, stating that "a more beautiful national park cannot be found.... A large number of prominent people in this State of Washington and Oregon are all interested in the establishment of an Olympic National Park" (Wickersham 1961:5). He submitted an article to *Century* magazine in April 1891, extolling the virtues of establishing such a park, but it was apparently never published. In July 1891 he wrote to Maj. J. Wesley Powell, then superintendent of the U.S. Geological Survey, including maps of the Olympic Peninsula and a copy of his unpublished article.[2] Wickersham's article was published long after his death (Wickersham 1961).

Upon completion of his 1890 explorations, Lt. Joseph O'Neil also suggested in a report (written in November 1890 and published in January 1896 as U.S. Senate Document 59 of the 1st Session of the 54th Congress) that the interior of the Olympic Peninsula would make a wonderful national park: "while the country on the outer slope of these mountains is valuable, the interior is worthless for all practicable purposes. It would, however, serve admirably for a national park. There are numerous elk—that noble animal so fast disappearing from this country—that should be protected" (O'Neil 1896:27).

The interior of the Olympic Peninsula was proclaimed the Olympic Forest Preserve in February 1897. The name of the reserve was changed to Olympic National Forest in March 1907 and to Mount Olympus National Monument on March 2, 1909, by proclamation of President Theodore Roosevelt, specifically to protect the elk found there (Mathias 1928; Roloff 1934). In October 1927 the monument was, by one account, "closed to hunting and trapping in perpetuity in an effort to conserve the wildlife of the mountains....however, supervised killing of the so-called predatory species was allowed" (Wood 1968:82). The U.S. Forest Service established the Olympic Primitive Area adjacent to Mount Olympus National Monument in 1936 (Wood 1968:84). Olympic National Park was established in June 1938 ("Olympic National Park Created" 1938), some thirteen years after mountain goats had been released on the peninsula, and the park was dedicated in June 1946. The history of the establishment of the park, which involves a conflict between conservationists and those with economic interests, especially the timber industry, is beyond the scope of this volume (see Hult 1954:213–19; McLeod 1984; Morgenroth 1991; Richardson 1968; Twight 1983).

Introduction of Mountain Goats to the Olympic Peninsula

In his book on the mammals of the Olympic Peninsula, E. B. Webster (1920:146), president of the Klahhane Club, writes: "While the fauna of the Olympic mountains has, for some peculiar reason, never included the

[2] These materials are still on file in the National Archives of the Smithsonian Institution in Washington, D.C.

Mountain Goat, it is the intention of the Klahhane Club (Olympic Mountaineers), aided by the Forest Service, to introduce the species as soon as it may be possible to secure them." Reports on the actual introduction are sketchy (see chapter 5). In one recent account, Olympic National Park researchers Bruce B. Moorhead and Victoria Stevens (1982:48) write:

> The first four goats were obtained from British Columbia. They were released January 1, 1925, along the northern edge of the Olympic Mountains near Lake Crescent (*Port Angeles Evening News*, January 2, 1925; Webster 1925). The record of other releases is less clear (Webster 1932). In either 1927 or 1929, seven or eight goats from Alaska were released, all but two again near Lake Crescent. The latter were apparently released 8 miles to the east, in the Elwha River valley near Baldy Ridge. A diary kept by one of the participants, Clallam County game warden Van Welch, is quoted later in a newspaper account (*Port Angeles Evening News*, Nov. 4, 1947). No other official records have been found. Welch indicated that all goat releases after 1925 occurred in 1929.

The references to articles written by Webster are not entirely accurate. For example, "Webster 1925" is a reprint of a newspaper article (probably *Port Angeles Evening News*, January 2, 1925) not clearly authored by Webster; it was reprinted in a biological journal with the final line, "From newspaper clippings sent in by E. B. Webster of Port Angeles" ("Introduction of Mountain Goats in the Olympic Mountains" 1925). Those reprinted newspaper clippings indicate that four mountain goats were "imported from the Canadian Selkirks ... at Banff" and released near Lake Crescent. The other article cited by Moorhead and Stevens (Webster 1932) begins with the statement: "Twelve goats have been liberated within the past few years at the foot of Storm King on Lake Crescent. One band of seven, a part of which were kids, has been seen; also a band of five."

In his 1946 monograph (commissioned by the NPS) on the mammals of Olympic National Park, Victor Scheffer (1946:129; see also Scheffer 1995:116–18) indicates that on January 1, 1925, four mountain goats (captured near Banff, but whether in Alberta or British Columbia is unclear) were released "on the shore of Lake Crescent at the foot of Mount Storm King" and that two additional goats were released "on the

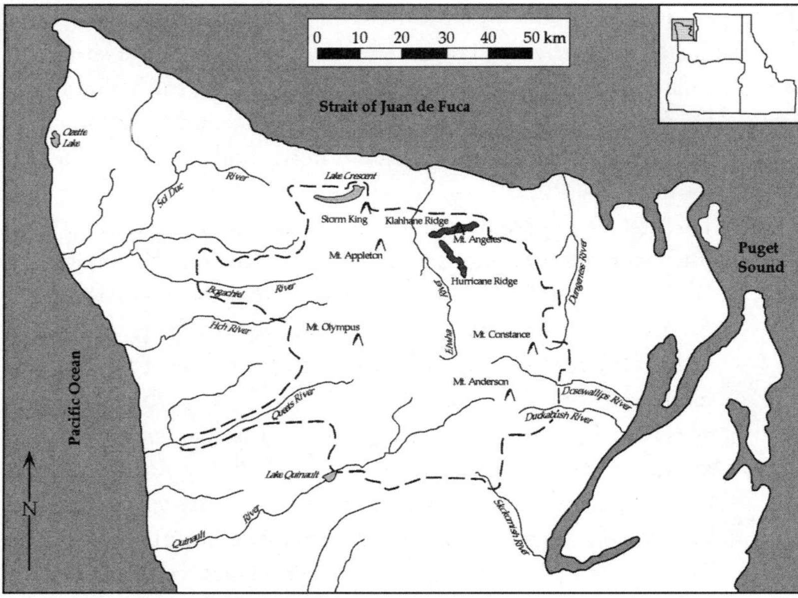

Figure 1.1. Major geographic features of the Olympic Peninsula. The boundary of the interior portion of Olympic National Park is indicated by the dotted line (the Pacific-coastal portion of the park is not shown).

Storm King Game Preserve" in 1929 (see fig. 1.1). According to Scheffer (1946:129), six more mountain goats were released on the Storm King Game Preserve in 1931. One report cited by Scheffer indicates that the first four were "three nannys and a billy," whereas another report indicates that they consisted of "two pairs." According to Scheffer, these first four goats were not seen again. The two goats released in 1929 came from the "Chugach Mountains near Cordova" (Scheffer 1946:129). The final six goats consisted of "three pairs" from "eastern Alaska" (Scheffer 1946:129). Moorhead and Stevens (1982) do not cite the information provided by Scheffer (1946), although they do cite his report for other information.

Victoria Stevens and Charles Driver (1978:165) indicate that "four individuals from the Selkirks in Alberta" were introduced in 1926, though I suspect that the date should read 1925. "During the next four years goats obtained from Alaska ... provided animals for two more introductions, approximately 11" (Stevens and Driver 1978:165). This makes it sound as if approximately fifteen goats were released between 1925 and 1929, inclusively.

Clarence Adams, administrative assistant of the U.S. Forest Service in Olympia, Washington, wrote in 1946 that he accompanied Chris Morgenroth (a forest ranger in Olympic National Forest) in the summer of "1923" when they released "two pair of mountain goats" on Storm King (cited in Morgenroth 1991:166). These are probably the same four individuals reported in "Introduction of Mountain Goats in the Olympic Mountains" (1925) and by Stevens and Driver as introduced in 1926: Adams (in Morgenroth 1991) and "Introduction of Mountain Goats in the Olympic Mountains" (1925) refer to one of the billies charging at Morgenroth. The important point here is that about a dozen mountain goats were released in the Olympic Mountains in the second half of the 1920s.

Post-1925 Dispersal of Mountain Goats

In 1929 a mountain goat was reported on Mount Appleton (Moorhead and Stevens 1982:49; see fig. 1.1), and in 1930 one goat was seen at Constance Pass, "about 30 miles airline from the planting site" (Scheffer 1946:130). Mountain goats were reported at the headwaters of the Dungeness River in 1931, on Mount Angeles in 1933, and on Mount Constance in 1935 (Moorhead and Stevens 1982:49). "Forest rangers estimated that there were 25 goats on the Olympic Forest" lands in 1937 (Scheffer 1946:130, 1995:118) or on lands that became Olympic National Park one year later.

The report of a mountain goat at the head of the Dungeness River (near Mount Constance) caught local journalist E. B. Webster's attention:

> I understand stories have come out from the east end of [Clallam] county concerning the presence in the upper Dungeness [River] country, of an old and very tame mountain goat. One story to the effect of a goat being lassoed by a Boy Scout may refer to this particular goat. It may be true, but why a goat should leave its fellows, and crossing the Elwaha [*sic*] River, travel twenty miles east on a low grass-covered ridge, when there were real mountains a much shorter distance to the south, would be a puzzle to me if I believed it. (Webster 1932:25)

What is intriguing about this statement is that Webster thought it unlikely that one of the transplanted mountain goats would have crossed the Elwha River and moved twenty miles east. Thus, he discounts this as a true record of a goat.

Moorhead and Stevens (1982) apparently find the record acceptable, but Webster doubted it. If there was a mountain goat at the headwaters of the Dungeness, was it one of the last *native* mountain goats in the Olympics? Webster considered the absence of native mountain goats from the Olympics "peculiar" and questioned the report of a mountain goat twenty miles away from where he knew the goats were. Some Olympic National Park personnel (e.g., Schultz 1993, 1994) find Webster to be a credible reporter. Yet Moorhead and Stevens (1982) do not mention Webster's questioning of this report. Is this because to doubt one of his statements they must also doubt another of his statements—that native mountain goats were absent from the Olympics in the early twentieth century?

Citing the admittedly scant newspaper accounts and park records, Moorhead and Stevens (1982:49) indicate that the transplanted goats (and their descendants?) dispersed east and south from Lake Crescent at a rate of about two miles a year, "reaching the east park boundary within 10 years [ca. 1938] and the south boundary in about 30 years [ca. 1960]." Goats were not reported from the Mount Olympus area until the late 1960s, but Moorhead and Stevens suggest that this might be a function of the few human visitors to the area.

Olympic Mountain Goat Population Trends

The introduced population of mountain goats consisted of about a dozen individuals. In 1937 there were an estimated twenty-five mountain goats (Scheffer 1946). Park biologist Bruce Moorhead estimated that the 1973 population consisted of 350 to 450 individuals (Olympic National Park 1987a) and that the population during the mid-1970s consisted of 1,000 individuals (Moorhead 1977; see also Olmsted 1979:1143). The 1980 population was estimated to be 500 to 700 individuals by Olympic National Park (1981) biologists and to be about 550 to 880 individuals by Victoria Stevens (1980; quoted in Olympic National Park 1982:5, 1983:5).

New populations always tend to grow quickly. Growth of the goat population probably was not restricted by predation because most large predators, such as wolves (*Canis lupus*), appear to have been locally extirpated prior to 1930 (Scheffer 1946:61; from 1927 on the Olympic Monument was apparently closed to hunting except for supervised killing of predators). Different counting methods were used during the 1970s and early 1980s; hence the variation in the estimates of population size (Houston and Stevens 1988). Census activities prior to the mid-1970s were few and largely incomplete (e.g., Jokerst 1960; Newman 1954) relative to later efforts.

Intensive aerial census activities in July 1983 resulted in an estimated population of about 1,175 animals (Houston et al. 1986; Olympic National Park 1984). The biologists who made the estimate stated that the census "indicates that the goat management problem facing [Olympic National Park] is very likely *larger* than anticipated initially" (Houston et al. 1986:136, emphasis added; see also Olympic National Park 1984:10). The estimated number of goats in 1983 served as a major catalyst for population-reduction efforts because it seemed that the mountain goats were modifying the Olympic ecosystem, and more goats meant greater and more rapid modification. Given the perception that the goats are "exotic" and management policies mandating that park managers recreate and maintain a pristine ecosystem, only one course of action was warranted: the mountain goat population of the late 1970s must be reduced quickly and significantly, if not eliminated. This is precisely what park personnel did.

Plants, Mountain Goats, and Ecosystems

In the 1970s and 1980s "research showed that over 50% of the [Olympic] Peninsula's rare plants occur in goat habitat along the park's eastern boundary" (*Interagency Goat Management Team* [*IGMT*] *Newsletter* 1:5; see fig. 1.2). Studies of those plants, including fencing some of them within what are called "exclosures," indicate that "goat numbers must be reduced by at least 80% before some native plant communities can appreciably recover" (*IGMT Newsletter* 1:5). Photographs taken with and without goats appear, in some cases, to be impressive examples of this phenomenon (Olympic National Park 1987a). These kinds of studies, in conjunction with the belief that mountain goats are exotic, have led park biologists to argue that "plant communities in the Olympic Mountains have evolved without goats" (Olympic National Park 1981:8) and to view the mountain goats as "an unwelcome addition to the evolved biota" (Houston et al. 1991a:87).

Olympic National Park biologist Ed Schreiner has been quoted as saying that the park consists of an ecosystem "that has evolved for thou-

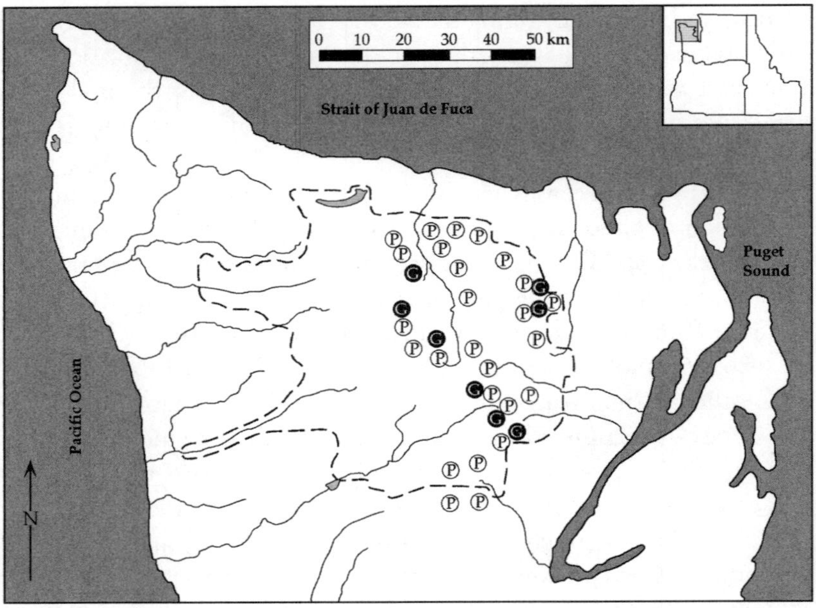

Figure 1.2. Map of mountain goat (G) and rare plant (P) concentrations on Olympic Peninsula as of late 1991. The boundary of the interior portion of Olympic National Park is indicated by the dotted line. Redrawn from *Interagency Goat Management Team Newsletter* 1.

sands and thousands of years, and what we've done by introducing a large herbivore [mountain goats] is upset all those natural balances" (Luoma 1993:41). This statement implies that Schreiner believes that mountain goats have not been present in the Olympics for several thousand years. It also implies that we have enough information about the Olympic alpine and subalpine ecosystem prior to the introduction of mountain goats in 1925 to construct an accurate model of what the ecosystem was like near the end of the nineteenth century. That ecosystem is, after all, the goal that must be sought according to park mandates. But as Stevens (1983:7–8) notes, the goats were "all but forgotten until decades [after their introduction], when their population levels were beginning to increase at a noticeable pace, and their habitat had become part of a national park." Thus, what we know of the century-old Olympic ecosystem in general, and the history of goats in particular, is actually very limited.

Secretary of the Interior J. A. Krug's (1946) remarks during the dedication of the park make it clear that the goats were in fact forgotten. Krug

(1946:549) indicates that, along with the Roosevelt elk, "favored as to food supply and the climate, flourished other animals *native to the region*, living their own lives protected from interference by man" (emphasis added).

THE CONTROVERSY

During the 1980s the goat population was reduced, and research was conducted to determine the effects of this reduction on the ecosystem. New plans were drafted in the mid-1980s that explored the ecological implications of major reductions in the goat population or complete eradication of that population. Statements that the mountain goats were exotic and were seriously disrupting the natural ecosystem of the park began to appear in scientific journals and popular magazines (e.g., Burger 1987; Hutchins and Stevens 1981; Moorhead and Stevens 1982). The possibility that goats might have been present in the Olympics in the past and became extinct prior to the historical period was not considered because the dogma was that they were exotic. Thus, the direction that future management efforts would take was set in 1981 when the park managers initiated an experimental removal program (Olympic National Park 1981, 1982).

Tests of contraception techniques were also performed at the request of the park in the 1980s. Personnel representing the Fund for Animals, local hunters and fishers, and other members of the public were asked to comment on potential plans (Carlquist 1990). Questions were raised by these groups: "Is the park rushing to an already-decided-on conclusion?" "Have all possible management options been explored?" "Has all of the relevant evidence been compiled, and does it clearly show that there is a problem?" The park failed to provide clear and concise answers; communication and agreement on issues quickly broke down. The resulting Olympic mountain-goat controversy concerns the efforts of park personnel to justify their plans to eradicate (Olympic National Park Superintendent Maureen Finnerty, October 8, 1994, *NBC Nightly News*; Olympic National Park 1995) or at least significantly reduce (Olympic National Park 1987a, 1987b) the local goat population and the efforts of others (e.g., Anunsen and Anunsen 1993) to slow, if not completely halt, the implementation of those plans.

On the one hand, we have park managers who operate under federal mandate to protect, preserve, and manage public lands based on the values assigned to those lands and the resources they contain. On the other hand, we have members of the public who seek to exploit those values but who may have different perceptions of what is valuable (e.g., Wagner 1989). The controversy concerns the issue of whether mountain goats should be allowed to stay in the Olympic Mountains, but in my view it hinges on the quality of scientific research and how it has been used to justify the position of one side or the other.

Wildlife management demands scientific research if its goals are to be attained and maintained. While scientific evidence changes through time as we learn more and build more complex models of how the world works, the scientific method itself is relatively stable (e.g., Murphy and Noon 1991). The values held by wildlife managers have changed over the years: goats were introduced by wildlife managers, who now seek to remove them. But the significance of the scientific method for the process has not changed. Nor have the definitions of native and exotic species, except that decisions should reside at the taxonomic level of subspecies, despite the wording of the definitions provided by the NPS.

By constantly stating that mountain goats are not native to the Olympics and that the ecosystem there developed without them, Olympic National Park personnel have shifted the burden of proof from themselves to other researchers, who must show that mountain goats *were* present during the last several thousand years (see also Hutchins 1995) or that the ecological impacts of today's goats are insignificant.

ABOUT THIS BOOK

Given the central role of scientific research in the controversy, my intent is to show that the management decision preferred by park personnel is founded on faulty logic, incomplete data, and misinterpretation. This does not necessarily mean that I believe that their preferred management action is the wrong one; in fact, the park personnel may have come to the correct conclusion, though I argue that it is much more weakly supported than they would have us believe.

By pointing out the fallacies of the reasoning involved, I hope to show how coming to the wrong conclusion might be avoided in similar situations in the future. In pointing out weaknesses, I must make some rather uncompromising evaluations of research performed by park scientists. These are not meant to be personal attacks. Some of my discussion may sound personal, but that is only because of the small number of individuals who generated the data and arguments I review.

I wrote this book because nearly every time I read a new article, news release, or report on the controversy a little research and thought convinced me that a serious error of omission or commission in data gathering, data analysis, or interpretation had occurred. Many of these errors, as documented in later chapters, are rather subtle and thus were easily—and typically—glossed over or ignored by park scientists. This makes it appear that they are hiding the truth under a series of white lies in order to strengthen their position that the mountain goats should be eradicated. Neglecting data that potentially contradict your favored hypothesis is, of course, a good way to convince people that your hypothesis is correct. But given the trust vested in the NPS by the nation's lawmakers and the public, surely such neglect must be deemed unacceptable.

Some will no doubt perceive that I am significantly less than objective in my review of the scientific research. My version of the controversy may, in some quarters, be equated with Alston Chase's (1987) remarks on the management of Yellowstone National Park.[3] Thus, I want to emphasize that I have never argued that mountain goats were present in the Olympic Mountains prior to their 1925 introduction. Nor, despite statements to the contrary by park biologists and others (e.g., Houston 1995; Scheffer 1993a), have I ever argued that the extant goats should be left alone if it can be shown that they were present prior to 1925. My main concern in writing this book is what I perceive to be the haphazard scientific research called upon by park personnel as justification for eradication of the goats.

It is my hope that other interested members of the public will be stimulated after reading this volume to ask questions of those public employees who are responsible for taking care of wilderness lands. It is critical that the public fully understand how and why wild places are being managed the way that they are. This book is the result of my own attempts to understand such issues.

Prior to 1988 the available literature repeatedly said that mountain goats were *never* native to the Olympic Mountains. My involvement began because I saw an interesting scientific problem. I came to believe that management of the goats was founded not on rigorous science but on a desire to meet the mandate of present NPS management policies: exotic

[3] Chase (1987) documents how the management of Yellowstone National Park has been founded in politics rather than based on science. The result, according to him, has been destruction of the Yellowstone ecosystem. As might be expected, Chase's study is controversial (e.g., Hess 1993), but has received some support from others (e.g., Kay 1990).

organisms are to be eliminated. In order to maintain a plan based on weak scientific research, public employees have produced still more research that is often biased or inconclusive. Yet they present the results to the public and call it science: the goats must go. I believe that the public has been misled by a strategy involving factoids and white lies founded on sloppy scientific research. This volume substantiates that view.

2

WILDLIFE MANAGEMENT, SCIENCE, AND THE NATIONAL PARK SERVICE

Wildlife management in the United States has a deep history. Many of its basic tenets were developed during a period of trial-and-error early in this century, when we were still learning about the structure and functioning of various ecosystems (e.g., Dunlap 1988). The role of the NPS in the development of procedures and goals of modern wildlife management has been documented in detail elsewhere (e.g., Sellars 1993; Stottlemyer 1981; Wagner et al. 1995; Wauer and Supernaugh 1983; Wright 1992). Many of the critical documents pertaining to the development of NPS wildlife management policies through 1992 have been reprinted (some in abbreviated form) in a single volume (Dilsaver 1994).

National parks present a special situation relative to the general landscape because they represent publicly owned lands and resources, and exploitation of wildlife (killing for food, fur, or sport) is strictly controlled or prohibited. This simple fact necessitates a unique set of rules and regulations (Wagner et al. 1995; Wright 1992).

BEGINNINGS

The 1916 Organic Act (or National Park Service Act), which effectively created the NPS, states that "the fundamental purpose of [national] parks is to conserve the scenery and the natural and historic objects and the wildlife therein and to provide for the enjoyment of the same in such a manner and by such means as will leave them unimpaired for the enjoyment of future generations." Early management of national-park lands was geared toward "preserving images and scenery" (Wagner et al. 1995:19). As Ervin Zube (1996:12) notes, "scenery was marketed [and] it was marketed with the eager cooperation of railroad companies that not only provided access to many of the parks in the pre-automobile era, but also provided luxury hotels and other creature comforts in parks for affluent American tourists."

Multiple use was practiced, and there was little effort to understand what we would today refer to as the ecosystems within a park (Sellars 1989, 1992). Maintaining scenery demanded no scientific research, but it did demand the enhancement of good animals (big game) and the elimination of bad animals (predators) and related activities such as the suppression of fires. This is today termed "facade management" (Sellars 1989:35).

The efforts of George M. Wright, a wealthy naturalist, in the late 1920s and early 1930s began to change what had gone before. Wright and biologists he hired began surveying the fauna in various national parks. He also initiated a series devoted to the publication of the results of his efforts. Much of the basis of modern wildlife management policies is found in his publications (Wright et al. 1933; Wright and Thompson 1935). For example, Wright emphasized that management actions should be preceded by scientific research, that native species that had been historically extirpated should be restored, and that exotic species that had become established within park boundaries should be eradicated. "The first step is to reconstruct the picture of the fauna of the area as it existed in its undisturbed environment before white men came" (Wright et al. 1933:9). The next step was to study human influences on the park and then conduct a complete survey of the fauna. The final step was to design a management plan that would "guide wild-life administration towards the general objective, which is to approach as nearly as possible the picture set up in the first step" (Wright et al. 1933:10).

The congressional act of 1872 that established Yellowstone, the first national park, had called for the preservation of the wonders of the area "in their natural condition" (quoted in Wagner et al. 1995:19). Wright proposed a way to do this and implied that "natural" meant anything pre-Euroamerican. Wagner et al. (1995:22) suggest that Wright's activities might have changed park wildlife management to a more science-based set of activities had it not been for his death in 1936, various political actions, and the Great Depression. Only in the late 1950s and early 1960s, with a national awakening of interest in environmental issues, did science-based management begin to replace management geared toward tourism and scenery.

Such management is beset with difficulties. One reason for this resides in the Organic Act of 1916 that created the NPS. As John Lemons

(1987:329) has noted, this "legislation is ambiguous regarding appropriate parks' values, forms of use and enjoyments, and concepts of naturalness." Nor does this legislation contain any indication of scientific standards or suggest whether or when visitor use or preservation of natural phenomena is to take precedence (Lemons 1987). Finally, the legislation "provides no standard of conduct or behavior against which the performances of park planners can be measured. Accordingly, management of parks' resources is by judgment and declaration" (Lemons 1987:330).

The concept "natural" and its related terms "pristine" and "primitive" are of greatest importance in the context of Olympic National Park's mountain goats. George Wright set the tone early with his statement that natural ecosystems were pre-Columbian. Although this position has been taken up by later researchers (e.g., Bonnicksen and Stone 1982, 1985), it is maintained only with difficulty, as we shall see.

The 1960s

A document generally referred to as the "Leopold Report" (Leopold et al. 1963), written at the request of Secretary of the Interior Stewart Udall, echoes the tone of G. M. Wright's early writings:

> As a primary goal, we would recommend that the biotic associations within each park be maintained, or where necessary recreated, as nearly as possible in the condition that prevailed when the area was first visited by the white man. A national park should represent a vignette of primitive America. (Leopold et al. 1963:32)
>
> The goal of managing the national parks and monuments should be to preserve, or where necessary to recreate, the ecologic scene as viewed by the first European visitors. (Leopold et al. 1963:43)

This is a laudable goal. The romance of "wilderness" is appealing to late-twentieth-century U.S. citizens, many of whom spend their day-to-day lives in traffic jams and industrial complexes (e.g., Bennett 1994). This goal cannot be attained, as the Leopold Report explicitly recognizes. But the authors persist: "Yet if the goal cannot be fully achieved it can be approached. A reasonable illusion of primitive America could be recreated, using the utmost in skill, judgment, and ecologic sensitivity. This in our opinion should be the objective of every national park and monument" (Leopold et al. 1963:32).

The goal of wildlife management outlined by the Leopold Report is unattainable for several reasons. It implies that only Euroamericans, who have been in North America a mere 500 years, have influenced or altered ecosystems. Native American peoples have been on the continent more than twenty times that long. The generally accepted date for the first peopling of the Americas is about 11,500 to 12,000 years ago, and people may have been in the Americas for 20,000 years or more (see Meltzer 1993). We must wonder if those peoples influenced ecosystems (e.g., Denevan 1992). Evidence suggests that they did (e.g., Bodmer et al. 1997; Kay 1994; Truett 1996), but "admitting that the Indian was an essential part of primitive ecology would mean accepting that, without the Indian, wilderness [is] an illusion" (Chase 1987:112). As Wagner et al. (1995:140) indicate, NPS statements on the concept of "natural" typically provide "no clear indication of whether aboriginal effects are included or not."

Benita Howell (1994:130) suggests that the NPS virtually since its inception has thought of natural ecosystems as "devoid of human contamination [and therefore] good [whereas] human intrusions on those ecosystems are bad." Modern park managers are taught this throughout their student and professional careers. This conception is "at odds with the understanding of cultural or human ecology developed since the 1930s in archaeology, cultural anthropology, and geography" (Howell 1994:130; see also Wagner et al. 1995:139–45). Two NPS biologists involved in study of the Olympic goats have rather cleverly avoided this problem by stating that the NPS "is primarily concerned with historic, post-Columbian species distributions" (Houston and Schreiner 1995:207).

Indian peoples have been so thoroughly enculturated or extirpated that most of the lifeways they followed prior to the appearance of the first European settlers have not existed for more than two centuries (see chapter 6). Their influence on ecosystems is often debatable, particularly when based on poorly preserved fossil and archaeological materials. In such cases it is easy to argue that "pristine conditions [can] be defined without [Indians]" (Chase 1987:114).

Even clear instances of Indian influences on ecosystems prior to the arrival of Euroamericans are easily ignored because of the difficulty of measuring them precisely with archaeological data. "Where land occupation was by native people and long-standing, and as the condition of the environment prior to their coming is relatively unknown, there is no option but to take that environment found upon arrival of Europeans as the natural one" (Hoerr 1993:32). In acknowledging our ignorance regarding pre-Euroamerican (for all intents and purposes prehistoric) ecosystems, this statement does not deny the influences of pre-Euroamerican peoples but renders such influences irrelevant because they are unknown.

To determine if these ecosystems that park managers desire are somehow natural requires conceptual gymnastics, such as determining "when human actions are in accord with nature" (Shrader-Frechette and McCoy 1995:138). As Karl Hess (1993:89) notes, "because the meaning of 'natural' is ambiguous and because it is left undefined in park policy, its application in the management regime ... provides a convenient way to escape accountability" (see also Lemons 1987:335–36). Too often we lack data on the immediately post-Columbian ecosystem; thus, "we are left in a quandary: how do we determine what is and what is not part of the natural scheme of things?" (Hess 1993:89). One searches in vain for a clear statement on what constitutes natural or native or pristine in a recent volume on management of U.S. national parks, although it contains many statements concerning disrupted, disturbed, and degraded ecosystems (Halvorson and Davis 1996).

Winfried Hoerr's (1993) statement quoted above indicates that we must have a source of information to construct a model of the ecosystem we are trying to recreate. Many would argue that the model must be quantifiable (e.g., Anderson 1991), but the lack of quantitative standards within the NPS makes it impossible to determine "how effectively the NPS is adhering to its mandate" (Lemons 1987:336). Given the limitations of our knowledge of prehistoric ecosystems and the notion that immediately pre-Columbian ecosystems are the goal, the ecosystem model must come from the earliest historical records. This is actually advantageous because we have learned that even recent prehistoric ecosystems were dramatically different from those of the present and even from those of the early historical period, regardless of the influences of people (e.g., Buechner 1953). Climates have fluctuated throughout the earth's history, and plant and animal communities have also changed and developed. Granting that Native Americans' impact on the ecosystems varied in kind and in magnitude over their long tenure, the goal of a single point in time seems reasonable.

The choice of the early historical period as the source of ecosystem models is, therefore, neither capricious nor impractical. It is virtually the only choice, *so long as it is realized that ecosystems continue to develop and change*—they are never static entities. The implications of this fact for wildlife management and modern attempts to preserve biological diversity are far-reaching (e.g., Graham 1988) but beyond the scope of this volume. A goal of maintaining a pristine ecosystem in perpetuity denies an earth history that has been and will continue to be dynamic, regardless of the presence or absence of humankind.

To approach the goals of creating and maintaining a pristine ecosystem, the Leopold Report recommended that "management be limited to native plants and animals. Exotics have intruded into nearly all of the parks but they need not be encouraged, even those that have interest or ecologic values of their own" (Leopold et al. 1963:34). Native plants and animals were to be restored to their ranges if they were absent. And "observable artificiality in any form must be minimized and obscured in every possible way" (Leopold et al. 1963:34). Finally, remembering that the goal "is to maintain or create the mood of wild America" and "preserve or restore a natural biotic scene" and that "above all other policies, the maintenance of naturalness should prevail" (Leopold et al. 1963:34, 35), the authors write that "the *first step* in park management is *historical research*, to ascertain as accurately as possible what plants and animals and biotic associations existed originally in each locality" (Leopold et al. 1963:36; emphasis added). This historical research provides the data for building a model of what immediately pre- or post-Columbian ecosystems looked like.

Another report issued in 1963 (generally referred to as the Robbins Report) stressed the importance of research in illuminating management goals and activities (Ackerman et al. 1963). But it also underscored that maintenance of an ecosystem in perpetuity "would not only be futile but contrary to nature" (quoted in Wagner et al. 1995:33). Thus, in the 1960s two things seemed clear. First, natural ecosystems were the goal, but the concept of "natural" was a commonsensical one and thus was ethnocentric and contingency bound. In other words, late-fifteenth-century environments (or at least those dating to the time of first Euroamerican contact) were natural. Second, active management was necessary to create and maintain a natural ecosystem; simply minimizing modern human impacts would not result in the attainment of desired goals.

WILDLIFE MANAGEMENT IN NATIONAL PARKS

U.S. wildlife-management policies have not been stable since the 1960s. As in any science or any management endeavor, information, perceptions, and values have changed, and thus wildlife management policies, procedures, and goals are not the same as they were thirty years ago. The history of wildlife management is intimately intertwined with an awakening public interest in environmental issues (Scheffer 1991) and, on national-park lands, with the history of the NPS (see Sellars 1992, 1993 for an introduction; and Wright 1992 and Wagner et al. 1995 for extended treatment). These histories have been strongly influenced by human "values that have the greatest currency" (Hoerr 1993:32), whether those values are scientific, economic, aesthetic, or humanistic (see Lemons 1987 for a brief but detailed overview of conflicting values; Matthews 1986 for a synopsis of legal issues; and Gray 1993 for extended treatment).

The history of wildlife management in the United States is full of examples of modifications to management policies and protocols to deal with changing perceptions of what is valuable (Dunlap 1988; Scheffer 1976). For example, early-twentieth-century eradication of pest species—especially large carnivores that preyed on large game animals or that might prey on domestic livestock—has been replaced by discussions of (re)introducing grizzly bears (*Ursus arctos*) and wolves to areas where they were historically eradicated, such as the Yellowstone ecosystem (Fischer 1995; Furtman 1993). We have learned much in the past several decades regarding how ecosystems work and change through time.

The 1970s

The NPS drafted a set of *Management Policies* in 1970 that were revised and updated in 1978 and again in 1988. The 1970 policies focused on maintenance of "natural processes of ecological succession" and the reestablishment of native life forms and "native environmental complexes," indicating that "appropriate management plans should be developed to control [nonnative species]" (USDI–NPS 1970). The structure and composition of natural ecosystems were to be determined "through historical and ecological research of plant-animal relationships."

The 1978 *Management Policies* provided definitions for "exotic species" and "native species" that served as the basis for many decisions about how plants and animals were to be managed. Those definitions, which are ultimately at the heart of the controversy surrounding the mountain goats found on Olympic National Park lands, are reproduced in full here:

> Exotic species are species that occur in a given place, area, or region as the result of direct or indirect, deliberate or accidental introduction of the species by humans ... native species are species which presently occur, or once did occur prior to some human influence, in a given place, area, or region as a result of ecological processes that operate and have operated without significant direct or indirect, deliberate or accidental alterations by humans ... direct or indirect, deliberate or accidental introductions by humans are ones that have permitted species to cross natural barriers to their dispersal capabilities thus giving those species opportunities to become established in areas previously inaccessible to them because of natural forces. (USDI–NPS 1978)

At first glance, the definitions seem reasonable, for one must have a scale by which to evaluate and measure phenomena, including naturalness (e.g., Anderson 1991). The definitions of exotic species and native species provide just such a scale for evaluation of the native or nonnative status of biological organisms found on park lands.

These definitions direct wildlife managers working on national parks toward management policies that coincide with the purpose of parks:

> Manipulation of population numbers of exotic plant and animal species, up to and including total eradication, will be undertaken whenever such species threaten protection or interpretation of resources being preserved in the park.... Control programs will most likely be taken against exotic species which have a high impact on protected park resources and where the program has a reasonable chance for successful control; programs are least likely to be initiated against exotic species which have almost no impact on park resources and where there is minimal probability for successful control. (USDI–NPS 1978)

Several key phrases in this quotation deserve emphasis. Control of exotic species can take the form of eradication *or control*, which I take to mean reduction of existing populations of the exotic species to reduce the impact on the occupied ecosystem. Control or eradication programs are to have a reasonable chance of success, which I take to mean that research will be done prior to the implementation of such programs in order to predict their probable success. Some biologists, however, would argue for elimination of exotics without antecedent research (e.g., Coblentz 1990; see also Bomford and O'Brien 1995 and Rainbolt and Coblentz 1997 for additional discussion). I would not disagree with such a protocol if it could be demonstrated beyond a reasonable doubt that the organism identified as exotic was, in fact, exotic.

The 1980s

Chapter 4 of the USDI–NPS *Management Policies* for 1988 begins with the statement that "the natural resource policies of the NPS are aimed at providing the American people with the opportunity to enjoy and benefit from natural environments evolving through natural processes minimally influenced by human actions" and notes that "the NPS will not seek to preserve natural systems in natural zones as though frozen at a given point in time." This recognizes the dynamic aspect of ecology and evolution and explicitly incorporates what had previously been a "missing element" of conservation efforts (Smith et al. 1993). Interference with natural processes in parks will only be allowed in specified situations, such as at the direction of Congress, in emergency situations, and "to restore native ecosystem functioning that has been disrupted by past or ongoing human activities." The term "natural" or "native ecosystem," while not defined explicitly, apparently means one that has not been or has only minimally been "disrupted by past or ongoing human activities."

Many factors complicate matters at the level of the park system. As the newest park management policies make clear, "wide variations exist in the degree to which the laws and proclamations creating the individual units of the national park system prohibit or mandate specific management actions.... [A]ll parks are complex mixtures of values and resources, each with its own unique qualities and purposes, each requiring specific treatment in the development and implementation of management strategies and operational plans" (USDI–NPS 1988). Thus, each park must have its own unique management policies and goals.

Another problem recognized in the 1988 policies is basic to modern park management. Those policies offer the directive to "assemble baseline inventory data describing the natural resources under [NPS] stewardship" (USDI–NPS 1988). In other words, the resources that are supposed to be managed are not known. The lack of data was made abundantly clear when Thomas Stohlgren and James Quinn (1992) evaluated baseline inventory data for forty national parks and monuments in the western United States. They found that larger parks tended to have more complete lists of plant and animal species within their boundaries than smaller parks; all parks, however, had mostly complete lists of mammals and birds but incomplete or no lists of many plants, reptiles, fish, and invertebrates. Perhaps this is not surprising, given that early management activities were aimed at making scenic mammals, for example, more visible to tourists than less scenic reptiles and invertebrates. But another study found that basic ecological data on ungulates in parks "indicated an important lack of information on population size and trends" (Robisch and Wright 1995:117). It is clear that we still have much to learn about the immigration to and extirpation of mammals in Mount Rainier National Park in the last 100 years (Dixon 1996; Weisbrod 1979; Weisbrod and Dragavon 1979). This seriously compromises the significance of studies such as William Newmark's (1995) attempt to detect overall trends in such histories in multiple parks.

The 1988 policies note that management of exotic species present on park lands should be undertaken; as in earlier policy statements, management activities can include eradication of exotic populations. Such action should take place when those exotics pose a threat to public safety, are damaging historical or prehistoric cultural remains, or are "interfering with natural processes and the perpetuation of natural features or native species (especially those that are endangered, threatened, or otherwise unique)":

> High priority will be given to the management of exotic species that have a substantial impact on park resources and that can reasonably be expected to be successfully controlled; lower priority will be given to exotic species that have almost no impact on park resources or that probably cannot be successfully controlled. The decision to initiate a management program will be based on existing, and where necessary newly acquired, scientific information that identifies the exotic status of the species, demonstrates its impact on park resources, and indicates alternative management methods and their probabilities of success. (USDI–NPS 1988)

In short, the issues that must be clearly addressed by scientific research are the exotic status of species and their impact on park resources.

Finally, the definitions of native and exotic animals in the 1988 policies are the same as in the 1970s policies:

> The native animal life is defined as all animal species that as a result of natural processes occur or occurred on lands now designated as a park. Any species that have moved onto park lands directly or indirectly as the result of human activities are not considered native.... Exotic species are those that occur in a given place as a result of direct or indirect, deliberate or accidental actions by humans (not including deliberate reintroductions).... The exotic species introduced because of such human action would not have evolved with the species native to the place in question and, therefore, would not be a natural component of the ecological system characteristic of that place. (USDI–NPS 1988)

The 1988 *Management Policies* outline how to recreate the native ecosystems. If a population of organisms that has disappeared from a park is to be restored, "subspecies used in restoration [must be the form that] most nearly approximates the extirpated subspecies or race"—in other words, be "the nearest living relatives of extirpated native species." The restoration is to occur only if "the species disappeared, or was substantially diminished, as a direct or indirect result of human-induced change to the species population or to the ecosystem." This part of the policy makes two important points. First, subspecies rather than species are to be used in restoration attempts. This makes biological sense (O'Brien and Mayr 1991), but it also underscores a weakness in the definitions of native and exotic *species.* Shouldn't these definitions concern subspecies too (Wagner et al. 1995:60)? Second, apparently only those (sub)species that were driven to extinction by human activity are to be restored to an area. This again raises the question of the influence of pre-European human occupation of North America. While the second point is generally dispensed with as irrelevant, the first is not.

Native or Exotic Subspecies and the Taxonomy of Mountain Goats

The definitions of native and exotic species found in NPS policy statements were still in use in the 1990s (e.g., Hester 1991). The wording of those definitions is contradictory, however. A species is native if it occurs or occurred (at some unspecified time in the past) in an area due to natural ecological processes. Yet a species is exotic if it is in an area as a result of human intervention. Thus, the definition of native species includes the dimension of time; as a result, "if mountain goats were ever prehistorically present in the Olympic Mountains then the statement that the species is 'exotic' would be false" (Lyman 1988a:21). In that article I suggested that discovery of evidence that mountain goats had been present during the last 1,000 years should prompt the NPS to "rethink policy issues" (Lyman 1988a:22). I later found out that Olympic National Park biologists were viewing the mountain goats in the park as nonnative from a taxonomic or genetic perspective.

Biologists today typically indicate that there are four subspecies of mountain goat in existence (e.g. Hall 1981; Rideout and Hoffmann 1975; Wigal and Coggins 1982). Ernest Booth (1947:598–99, 602–3) indicates that the subspecies *O. a. americanus* (found in the Cascade Mountains of Washington) and *O. a. missoulae* (rarely found in northeastern Washington) are native to Washington, whereas *O. a. columbiae* and *O. a. kennedyi* were introduced to the Olympic Peninsula in 1924 (1925?) and 1929–30, respectively. Booth (1947:603) speculates that Webster's (1932) two groups of seven and five goats represented *O. a. kennedyi* and *O. a. columbiae*, respectively.

> Apparently the two races [subspecies] of goats do not mix, but maintain their own separate groups even though occupying the same general territory. The future of this population can hardly be determined at present. The goats seem to be able to thrive in the Olympic Mountains, but what the outcome will be since two subspecies have been liberated in the same region can hardly be predicted. They may maintain separate populations, or they may interbreed. (Booth 1947:603)

We do not know if the two introduced goat subspecies interbred (though that seems possible), nor do we know the accuracy of Booth's speculation regarding the subspecific identity of the two groups of goats reported by Webster. Scheffer (1946:129) reports that the "fate of the first [1925] planting is uncertain" as no goats were listed in the Forest Service's annual wildlife reports for the years from 1925 through 1929. Perhaps the first transplanted goats all died without leaving offspring. But the ultimate fate of the two subspecies on the peninsula is, as it turns out, probably irrelevant to the controversy because biologists cannot agree today if there really are distinct subspecies of mountain goats. Based on a 25-year-old study (Cowan and McCrory 1970), "subspecific categories remain to be confirmed or abandoned based on additional scientific data for this species" (Rideout 1978:149).

This means that even if NPS policies are rewritten so that native and exotic species become native and exotic subspecies it will still be unclear if the subspecies (or hybrid of two subspecies) of mountain goat presently occupying Olympic National Park is exotic. The biological community must agree, on the basis of empirical data, that the mountain goats transplanted to the Olympics were a nonnative form. Then it must be established that mountain goats never occurred there prior to 1925 or, if they did, that humans had no hand in their local extinction.

But subspecies designations may also be irrelevant. Recall that reintroduced organisms should be "the nearest living relatives of extirpated native species" (USDI–NPS 1988). What does this mean? The International Union for the Conservation of Nature and Natural Resources (IUCN), now called the World Conservation Union, drafted guidelines for the reintroduction of animals to their original range that are relevant. Reintroduced taxa "should be of the same taxonomic unit (and ideally closely related genetically) as those which were extirpated....Release stock ideally should be closely-related genetically to the original native stock" ("Draft Guidelines for Reintroductions" 1992:2–3). Subspecific designation is less important than genetic relatedness. Thus, while begging the question of how close genetic relations should be, the IUCN policies are more lucid than those of the NPS. This also clarifies why NPS might view the goats in Olympic National Park as exotic—having come from Canada, they are the wrong genetic form. Apparently, if goats from Mount Rainier (the typically suspected source of prehistoric goats in the Olympics, if any in fact existed) had been transplanted to the Olympics in 1925, then this argument would lose much of its force.

THE CENTRAL ISSUE

The role of scientific research is abundantly clear, given G. M. Wright's writings and the Leopold and Robbins reports. To manage nature, we must understand the structure and functioning of ecosystems. But we are constantly learning, so we could argue that we never know enough to make the best management decisions possible. Emotional issues have come to play an increasingly important role in wildlife management (e.g., Scheffer 1974; Wagner 1989), no doubt because of the public's growing awareness of the concept of ecology (Scheffer 1991). Public views "are rooted in emotion as well as reason and, because they are not purely rational, they disturb many practicing wildlife biologists" (Scheffer 1976:51). As noted in the USDI–NPS 1988 *Management Policies,* "national interest must dictate all decisions affecting public or private enterprise in the parks." Thus, the public has a legal right to comment on management activities. Interestingly, a 1995 poll of residents of Washington state found that over 70% of the respondents favored leaving Olympic goats alone (R. Anunsen, personal communication, October 31, 1995).

Wildlife manager James Bailey (1982) spoke to the tactical problems involved in making management decisions. Wildlife managers must have "comprehensive, quantitative, and continuous knowledge of the [eco]system they manage"; "a clear definition of publicly agreed upon management goals"; and full knowledge and understanding of the effects of all possible management activities. Bailey (1982:363) concluded that "if wildlife biologists could amass this information, the rest of their job would be simple. They would merely select the best set of treatments [management activities] for achieving the clearly defined goals."

But reality is not so simple, given the complexities of wildlife and ecosystems, limited budgets and limited time for undertaking the requisite research, and a diversity of conflicting goals of management (Bailey 1982). This is especially true on national-park lands (Houston 1971; Lemons 1987; Wright 1992). Thus, wildlife managers "muddle through," according to Bailey (1982), making the best decisions possible based on the information available at the time. Management plans are constantly being modified in light of new information (see also Walters 1986). Other wildlife managers agree (e.g., Dorrance 1983; Gavin 1989, 1991; Macnab 1983; Murphy and Noon 1991; Nudds and Morrison 1991; Romesburg 1981, 1989; Sinclair 1991; Wright 1992).

Some argue that one reason that wildlife managers have to "muddle through" is chronic underfunding (Scheffer 1991; see also Labisky et al. 1986). Pleas that the parks be closed to the public were made in the 1950s, and in the 1980s it was reported that parks were "still suffering from neglect" (Scheffer 1991:41). The NPS asked Congress for more money early in the 1980s in an attempt to slow the destruction and degradation of the lands and resources under its control, but the clearly nonconservation mindset of the Reagan administration (1981–88) and its initial appointments to head the Department of the Interior and the Environmental Protection Agency probably squelched any hope for a thorough research program that would serve as a basis for sound management (Scheffer 1991:178–79). Others have argued that the dual purposes of the NPS have resulted in the majority of funds being spent on tourism-related matters at the expense of research and protection of natural resources (e.g., Wagner et al. 1995:88–89; see also Davis and Halvorson 1996:6).

The central issue in the Olympic mountain-goat controversy hinges on the definitions of native species and exotic species. Biologists involved in the Olympic goat controversy have suggested that only post-Columbian ecosystems are of concern (Houston and Schreiner 1995), although the policies do not formally state this. Wagner et al. (1995: 171–73), however, suggest that pre-Columbian or at least preindustrial ecosystems are of greatest concern to NPS personnel. Obviously, the alternative chosen will result in 300-year-old mountain-goat remains from the Olympic Peninsula having quite different significance than 600-year-old remains.

The mountain goats presently found in the Olympics can be considered exotic only if they were never present prior to their artificial introduction (or at least were not present between 1492 and 1925) *or* if native goats became extinct prior to 1925 for some natural reason. The controversy therefore hinges on two questions. First, did goats ever occur in the Olympic Mountains prior to 1925? If we can establish an affirmative answer to this question, particularly if goats were present sometime between 1492 and 1925, then we must answer the second question: What caused the native goats to become locally extinct? If we can show that human hunting had something to do with that local extinction then the extant mountain goats, *by definition*, are native.

One might argue, as Chris Maser (1990:131) does, that "nativeness" is a matter of degree: "the further into the past that one can trace his or her ancestry in a given place the more native he or she is thought to be." Such a view would suggest that the goats in Olympic National Park are native, given their 70-year tenure there, but less native than, say, the plants. Olympic National Park (1981:8) indicates that "plant communities in the Olympic Mountains have evolved without goats." This is the argument that links the definition of mountain goats as exotic and the park's efforts to remove them. The mountain goats are "an unwelcome addition to the evolved biota" (Houston, Schreiner, Moorhead, and Olson 1991:87). In order to preserve the native plants, the exotic goats must be eliminated.

SCIENCE

Science is simply a way of knowing. This does not mean that all scientists agree on what the best scientific way of knowing is (compare, for example, Matter and Mannan 1989 and Romesburg 1989) or even on what is thought to be known. But the nifty aspect of science is that it is self-correcting. Mistakes have a rather short life span. A hypothesis is a testable statement that "describes how we suppose [a] bit of the world works" (Sinclair 1991:767). Hypotheses are continually being tested and refined because "if we assume a hypothesis is correct when, in fact, it is not, then a [wildlife] management decision based upon it also will be wrong" (Sinclair 1991:768). But sometimes when a hypothesis has been around for a long time it is assumed to be true. "By frequent reiteration of the hypothesis it appears 'true,' and so there is no longer any [perceived] need to test it. At this point the hypothesis has been transformed into dogma" (Sinclair 1991:768) or a factoid. The hypothesis that mountain goats never naturally occupied the Olympic Mountains was dogma in the 1970s.

Wildlife managers often test biological hypotheses to find out the potential result of altering one variable in the ecosystem they are studying (Sinclair 1991). They need to know how the variables of interest interrelate prior to performing their experiment, during the experiment, after the experiment, and in control samples or samples to which a protocol is not applied for comparative purposes.

Testing multiple hypotheses ensures that a hypothesis is not chosen simply because it is favored and, more importantly, avoids the problem of unfalsifiable (and thus potentially incorrect) hypotheses becoming dogma. A related technique involves testing a null hypothesis—one that is written in such a manner as to be capable of being nullified or shown to be false. Falsifying the null hypothesis implies that there is an alternative hypothesis. Hypotheses are tested with data. Data are merely observations of the world and serve as evidence only when they confirm or negate a hypothesis.

Some scientists argue that scientific research belongs to the public: "Research cannot be considered complete until its findings are presented in the literature. Thus if the community (scientific, field-naturalist, political, etc.) is to profit from observations and the results of research, these must be published" (Smith 1976:101). This position derives from the notion that "one of the aims of scientific journals is to provide the media in which controversial matters can be debated" (Webster and Yaalon 1994:10). Thus, some scientists argue that "the output of a scientist must be put on public display before it can be considered a scientific contribution" (McCain and Segal 1969:45) because this permits independent peer review. The peer-review process allows an editor to call on experts to evaluate papers and, ideally, ensures that only research of the highest quality will be published (Frasier 1988). "To be accepted by one's peers is naturally preferable to acceptance by a group of people drawn at random and, conversely, to be rejected by them is doubly hurtful" (Maddox 1989:11).

Other scientists suggest that "reasonable people do not contest the data or request peer-reviewed papers to determine [scientific—in this case, wildlife management—issues]" when published data are not contestable (Hess 1993:134). The 1988 USDI–NPS *Management Policies* state: "The results of scientific activities conducted in the parks will be made available to park managers, the scientific community, and the public through both technical publications and the popular media." Olympic National Park resource management specialist Paul Crawford (1993:15) stated in a quarterly journal published by the NPS that all research on the mountain goats between 1978 and 1993 had been or would be peer reviewed. As we will see in later chapters, that statement is not true.

The importance of scientific research in the management of national parks has become abundantly clear in the last several years. For example, the National Research Council (1992) called for better science on the part of the NPS. The council argued that the new structure it recommended should "stress science in the national park system and guarantee long-term financial, intellectual, and administrative support" (quoted in Dilsaver 1994:446). Key elements in the recommendations are that "there must be efforts to enhance the credibility and quality control of the science program" and that this requires "explicit legislative mandate for a research mission of the NPS" and "separate funding and reporting autonomy" for the science program (in Dilsaver 1994:446). The new science program was, in part, to "meet the needs of resource managers." The NPS (1993) adopted these notions, and one result was the formation of the National Biological Survey. The NPS noted that "resource management decisions should be presented as hypotheses" (NPS 1993:1) and that attempts should be made "to help overcome the image that national parks are not receptive to outside research" (NPS 1993:6). Thus, perhaps some of the problem in the Olympic National Park mountain-goat controversy will be avoided in the future. This, however, remains to be seen.

THE OLYMPIC NATIONAL PARK MOUNTAIN-GOAT CONTROVERSY

Since the early 1980s, Olympic National Park biologists have argued that the mountain goats in the park are exotic. They have stated that the goats compete for food with the Olympic marmot (*Marmota olympus*) (Burger 1987), they trample and feed on rare plant species (e.g., Pfitsch et al. 1983), they trample mountain beaver (*Aplodontia rufa*) burrows and compete with them for forage (Chadwick 1983:176), and their wallows exacerbate erosion and disrupt plant growth (Olympic National Park 1987a). Park biologists reason that because the plant species, the other animal species, and the geology are all native to the Olympic alpine ecosystem, the goats must go. Of course, if it can be shown that the mountain goats are also native, then they must—by the same reasoning—be allowed to stay.

The controversy surrounding the Olympic goats has served as an example in two recent books on wildlife management in national parks (Wagner et al. 1995; Wright 1992). In one of these books, four pages of a twenty-page chapter entitled "The Management of Alien Animals" are devoted to the Olympic goats (Wright 1992:101–5). This discussion begins with the following statements:

> For years the National Park Service has struggled to develop a workable plan to eliminate the mountain goat from Olympic National Park. This effort epitomizes all of the problems that alien species can cause in parks and the difficulties the Park Service has in explaining and carrying out its policies.
>
> Mountain goats are native to many of the mountainous areas of northwestern North America, but because of the isolation of the Olympic Mountains from the Canadian Rockies and the Cascade Range, mountain goats never colonized the area. Surveys of the region in 1897 and 1899 by eminent biologists found no evidence that mountain goats existed in the Olympics. (Wright 1992:101–2)

The message is that the Olympic Mountains are presently biogeographically isolated from various alpine-adapted mammals found elsewhere, so these species were never in the Olympics in the past.

Gerald Wright (1992:102) goes on to note that "concern that the goats were damaging subalpine vegetation caused [the NPS] to initiate studies of the mountain goats in the seventies. Research throughout the seventies and eighties examined the impact of the mountain goat on plant communities and soils.... There was concern that some species of unique endemic plants would eventually be eliminated." In Wright's (1992:103) view, the controversy resides in the fact that the NPS views the mountain goats as an exotic species that does not belong in Olympic National Park, whereas at least some members of the public argue that mountain goats are "a very popular animal"; that the "ecological damage the goats are causing is not readily apparent to the casual observer"; and that elimination of the park's goats will "ultimately eliminate the sport harvest outside the park" that is administered by the Washington Department of Wildlife and Olympic National Forest.

In my view, the research on the exotic status of the mountain goats and on their influence on the alpine ecosystem has not been exhaustive and rigorous. Some of the park biologists' hypotheses have not been subjected to thorough testing. But many of these hypotheses are regularly used as warrants for arguing that the mountain goats should be eradicated.

3
POPULATIONS

The argument that the mountain goats in Olympic National Park must be eradicated hinges in part on the apparent negative influences they are having on the alpine and subalpine ecosystems. Because such ecological effects are density dependent, discovering that there is a large population would suggest a more serious ecological threat than determining that the population is small.

TERMS, CONCEPTS, AND METHODS OF POPULATION ESTIMATION

A "census" is an actual count of organisms in an area. But because wildlife species do not fill out questionnaires, estimates of the population tend to be the rule. Deriving an "estimate" of a population typically involves sampling the population and then calculating a statistical approximation of the size of the entire population based on the size of the sample. Both the sampling procedures and the calculations can take several forms, but they typically involve the use of ratios of the number of organisms seen per unit of time spent surveying or the number of organisms tallied per unit of geographic space examined. Thus, for example, if one sees 10 organisms in a one-hectare sample of a ten-hectare area, then the ratio of 10 organisms per hectare is used to estimate that the total population is 100 organisms (10 organisms per hectare × 10 hectares). As David Davis and Ray Winstead (1980:223) emphasize, the critical assumption here is that the ratio observed in the sample is the same as the actual ratio of the population throughout the area of interest; they note that "the big job of the wildlife biologist is to test this assumption." Also, if the biologist takes more than one sample, then "the assumption must be made that the same proportion of the total population is tabulated each time the same [counting] technique is used under similar field conditions with the same species" (Davis and Winstead 1980:224).

Techniques for estimating wildlife populations have seldom been checked against populations of known size (Davis and Winstead 1980: 244). Perhaps that is to be expected given that estimates are typically taken in order to discover the size of a poorly known population. The net result is that the appearance of accuracy provided by sophisticated statistical methods in calculating estimates may be illusory. While no statistical assumptions may have been violated, nonstatistical assumptions such as the uniformity of the samples throughout an area may be false. Nonetheless, as Davis and Winstead (1980:245) make clear, even a rather gross level of accuracy in population estimates is often sufficient for wildlife managers, who are typically unimpressed by management techniques that will increase or decrease populations by (an undetectable) 5%; "to be worth the effort, the management technique must have a large effect [and] fortunately even a [population estimation] method of low precision can detect a large effect" (Davis and Winstead 1980:245). Thus, if the *IGMT Newsletter* (1:5) is correct in asserting that the Olympic mountain goat population "must be reduced by at least 80% before some native plant communities can appreciably recover," then such a reduction would be easily measured with available population estimation techniques. Similarly, as park biologist Ed Schreiner argues, "if we truly want complete recovery [of plant communities], we can't have any goats" (quoted in Burger 1987:29); a 100% reduction in goat numbers would be easily detected.

THE 1983 POPULATION ESTIMATE

Part of the Olympic mountain-goat problem resides in the apparent explosion of the goat population. Based on their estimate that the 1983 goat population numbered about 1,175 individuals, park biologists concluded that "the goat management problem facing the National Park Service is very likely larger than anticipated initially" (Houston et al. 1986:136). Except for the apparently unacceptable estimation of 1,000 goats by Bruce Moorhead (1977), previous estimates were less than 1,000 goats. The magnitude of the suspected population explosion is founded on somewhat less than rigorous estimates of the pre-1983 population size, however. In some cases, park biologists have attempted to estimate past population sizes from field notes of other researchers, which typically do

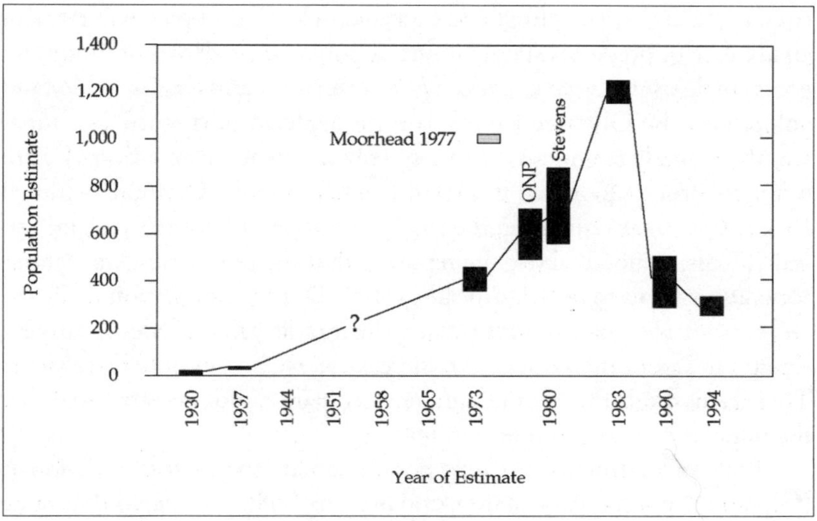

Figure 3.1. Mountain goat-population trends in the Olympic Mountains. Note that Moorhead's (1977) estimate of 1,000 individuals does not seem to be generally accepted and that two estimates of population size are available for 1980 (one by park biologists, one by Stevens).

not allow statistically rigorous estimation (e.g., Houston and Stevens 1988; Olympic National Park 1984). If those pre-1983 estimates are too small, then they indicate a larger than actual population increase in the early 1980s; if the 1983 estimate is too large, the same effect results. There is, however, no apparent way to test the earlier estimates of population size.

Several aspects of the reported recent population explosion warrant comment. Available population estimates are summarized graphically in figure 3.1; the estimate of 1,000 individuals for the 1976 population by Moorhead (1977; quoted by Olmsted 1979 and Stevens and Driver 1978) is set off from the other estimates because it is only mentioned three times in the literature I have examined. The 1991 Resources Management Plan prepared by Olympic National Park indicates in one place that the estimated goat population was 1,200 in 1988 and in another place that there were "about 1000 animals in 1983." The latter estimate seems to be wrong (Houston et al. 1986); the former estimate is probably based on the 1983 estimate and is probably wrong for reasons that will become clear.

Available estimates indicate that the goat population increased between 1930 and 1983, as might be expected in an unoccupied habitat with an absence of predators. Populations of potential predators of mountain goats such as wolves, cougars (*Felis concolor*), and eagles (bald and golden) on the Olympic Peninsula were depleted during the late nineteenth and early twentieth centuries; wolves were totally eradicated. Park biologist Bruce Moorhead is quoted in a park report (Olympic National Park 1987a:20) as estimating the 1973 population at between 350 and 450 individuals; as noted above, he reported that the population was "about 1000 animals" in 1976 (Moorhead 1977:2). Did the population really increase from 400 goats in 1973 (taking the middle value of Moorhead's estimate) to 1,000 individuals—an increase of 250%—in only three years? That seems doubtful, because unidentified park biologists estimated that the population was about 600 in 1980–81.

Biologists estimated the 1980 population at "500 to 700" individuals (Olympic National Park 1981:5), and Stevens (1980) estimated that same 1980 population to consist of "554 to 884" individuals. Clearly, these numbers represent *estimates*; hence the discrepancy. Curiously, subsequent reports by park biologists quoted the higher of the two estimates and did not mention the lower one (Olympic National Park 1982:5, 1983:5). Were Stevens's estimates methodologically sounder? Were park service estimates less rigorous statistically? We are not told. All we can conclude is either that one (or both) of the estimates is seriously inaccurate or that the goat population suddenly increased by 50 to 184 individuals in 1980.

An aerial census by Olympic National Park biologists in 1983 estimated the mountain goat population to number about 1,175 individuals (Houston et al. 1986). That estimate was soon reported in the widely circulated popular literature as "almost 1200" (Burger 1987:28). Clearly, if the goats' impact on the alpine ecosystem was dependent on population size, then larger populations would have greater ecological impact. Was the larger estimate in 1980 chosen simply because it was larger? Was Moorhead's (1977) estimate of a population of 1,000 goats given to Olmsted (1979) because she was studying the impact of goats on alpine vegetation and more goats mean more impact?

Why would the population grow from about 400 individuals in 1973 to nearly three times that many in ten years?[1] Such an increase is all the

[1] Several biologists I have spoken with have posed that question to me, including one who worked extensively with the elk of the Olympic Peninsula in the 1980s. They asked because they did not know the answer rather than to determine if I knew the answer.

more impressive because fifty-two goats were removed from the park in the summer of 1981 by park biologists (see Johnson and Moorhead 1982 for a description of the techniques used); in 1982 fifty-one additional goats were removed; and in 1983 fifty-two more goats were removed *prior to taking the census* (Olympic National Park 1984:15). If the 1973 and 1983 estimates are correct, then clearly the population had exploded. The park biologists argue that the population would have decreased some unknown amount as a result of the more "normal" winter snowpack in the winter of 1982–83 (Houston, Moorhead, and Olson 1991). This, too, suggests that the pre-1983 population may have been larger than indicated by the summer 1983 estimate.

Estimation Procedures

But is the 1983 estimate correct or at least accurate to some specifiable degree? The estimate of 1,175 mountain goats was obtained by aerial census activities in July 1983. Park biologists actually observed a total of 396 individual goats. The population estimate of 1,175 was derived by assuming a 66% chance of sighting an individual goat and accounting for the fact that not all places where goats were known or suspected to occur were inspected. This resulted in the multiplication of the observed goat population by the reciprocal of the estimated chance a goat would be seen. Thus, if 10 goats were seen in a place, assuming that only 66% of the goats actually present there were seen, those 10 goats would be multiplied by 1.52 to obtain an estimate of 15.2 goats in the area if sightability had been 100%. The second step in the estimation procedure involved multiplying the average density of goats (number of goats per unit of land area given 100% sightability) by the total land area suspected to be inhabited by goats (Houston et al. 1986; Olympic National Park 1984).

In fact, this two-step procedure was only part of the method used to produce the estimate of 1,175 goats. It was applied in areas where goats were suspected to occur in medium and low densities. In areas where goats were suspected to occur in high densities, total counts (actual censuses) were attempted. And the goat population in a fourth area was calculated using an index. The two-step estimation procedure was used for nearly 86% of the acreage examined (42,995 hectares out of 50,063 hectares) and resulted in 78% of the estimated population (917 of 1,175). Thus, I focus on those areas and that procedure here.

First, on a rather pedantic note, I found four errors in the appendix to the unpublished report that describes the population estimation procedures (Olympic National Park 1984:134–35). Three of these are simply errors in describing the variables in the formulae used to calculate the population estimates. The fourth is an error in listing the value of one variable in one formula. All four of these are apparently simply typographical errors because by working through the mathematics I can replicate most (but not all) of the values the park biologists present. Unfortunately, not all of the relevant data are presented. It is not always possible to publish all details of a data set; summaries of the most relevant data are typically preferred by editors of scientific journals. But the omission of relevant data from an unpublished in-house park service report is inexcusable. It effectively precludes evaluation of the unpublished report and of conclusions derived from that report published in scientific journals. Failure of the park biologists to include all relevant data leaves their conclusions open to question.

The published version of this report (Houston et al. 1986) is a summary of the relevant data; some critical data are not provided, so the mathematics performed by the biologists cannot be replicated. Some variables are undefined; thus some of the reported values for those variables seem to appear from thin air. Again, I found at least one error in the data presented. The potential significance of the missing data is clear when we review the estimation procedures.

Park biologists divided the habitats frequented by goats into four kinds of areas. As noted above, I am concerned here only with areas where goats were thought to occur in low and medium densities. Relevant data are summarized in table 3.1. The term "stratum" simply means part of a population or subpopulation (in this case, a subpopulation of the areas where goats occur), and "sampling" signifies that only part of each stratum was actually examined. The areal units into which each stratum were divided are termed "blocks"; some blocks in each stratum were examined.

Table 3.1 shows that 282 mountain goats were observed in the blocks examined in the two strata. Only 66% of the goats present were thought to have been seen, so it was estimated that there were probably about ([1.52 × 277 =] 421 + 7.6 [= 1.52 × 5] =) 428 or 429 goats present in those examined areas. The average density of goats, or the frequency of goats per unit of area (hectare), was calculated for each stratum. Thus, for the low-density stratum, this was calculated to be (7.6 goats ÷ 2,847 exam-

TABLE 3.1.
Data for Estimating Mountain-Goat Populations in Low- and Medium-Density Strata (from Olympic National Park 1984 and Houston et al. 1986)

Sampling stratum	Total area in stratum (hectares)	Total number of blocks	Area examined (hectares)	Number of blocks examined	Number of goats observed	Adjusted number of goats[a]
Low density	13,135	26	2,847	6	5	7.6
Medium density	29,860	59	14,251	28	277	421

[a] This is the number of goats observed multiplied by the reciprocal of 66% sightability, or 1.52.

ined hectares =) 0.00267 goats per hectare, multiplied by the total area for the low-density stratum to obtain an estimate of 35 goats (= 0.00267 goats per hectare × 13,135 hectares). Similarly, for the medium-density stratum, the density of goats was calculated to be (421 ÷ 14,251 =) 0.02954 per hectare, multiplied by the total area of the medium-density stratum to obtain an estimate of 882 goats (= 0.02954 × 29,860).

A number of questions can be asked about this two-step procedure. First, is 66% sightability reasonable? In a review of the historical evidence for the pre-1925 presence of mountain goats, Olympic National Park historian Susan Schultz (1993:53; 1994:291) writes: "Mountain goats are large, conspicuously white animals. Nineteenth-century observers reported that they could be seen 'from a great distance feeding in large droves' where they were present," citing an 1860 report written by George Suckley and George Gibbs. They were quoting "an old Indian trader, and at present United States Indian agent" living in west-central Idaho who told them that mountain goats were "found in *great numbers* on the bald hills and bare mountains of the locality, and that upon these they can be seen from a great distance feeding in 'large droves'" (Suckley and Gibbs 1860:137; emphasis in original). Schultz suggests that if mountain goats had been present, given their supposed high visibility, they would have been seen.

Schultz argues that mountain goats are "conspicuous" because they are white. This obscures the fact that the goats often lie on snowbanks, particularly in the summer, as a way to regulate body temperature. White can be a camouflage color, particularly in the winter (Chadwick 1983 titled his book *A Beast the Color of Winter*). Also, while it is true that goats are easily seen when they occur in "large droves" against a nonsnow background, they would be less visible singly or in small groups, especially in a snow-covered landscape. Park biologist Bruce Moorhead (1977:2) notes that "mountain goats are aptly described as 'crotchety loners,' with only limited tolerance for their own kind. Although temporary feeding groups of 20 to 40 or more females and young do occur in summer and fall on optimal range, the average group size is more like two or three." Finally, the nineteenth-century report cited by Schultz indicates that mountain goats were easily seen when they were "on the bald hills and bare mountains," not when they were in timber.

Seeing all of the mountain goats in an area is typically thought to be improbable. But are 66% of them sightable? Joanne Michalovic (1984:138) has suggested that "there has been no one systematic methodology formulated for censusing mountain goat populations that is applicable to all circumstances." She used a combination of pedestrian ground surveys and aerial counts to estimate the abundance of mountain goats in Mount Rainier National Park. Following others, such as Stevens (1982, 1983) in the Olympics, Michalovic (1984:148) concluded that about 50% to 70% of the goats present in an area can be seen by ground surveys. She found that counts from aerial surveys were consistently less than counts from ground surveys in several areas of Mount Rainier National Park. Thus, perhaps as Houston et al. (1986:135) suggest, their "estimate of 66% observability [from the air] is conservative." For example, Houston and Stevens (1988) report that a 1959–60 helicopter census of mountain goats on Klahhane Ridge suggested 59% sightability; a 1964 fixed-wing aircraft census of that same area suggested 44% sightability (see also Olympic National Park 1984). Rolf Johnson (1983) estimated 63% sightability during an aerial census in the northeastern corner of the Olympic Peninsula. Thus, several other researchers would multiply the number of goats observed by a larger number than 1.52. It seems, then, that the 66% sightability figure is reasonable, but it lacks great precision.

The 66% observability factor was calculated using the average of two counts of goats on Klahhane Ridge taken twenty-two days apart in the summer of 1981 and a pair of formulae for estimating the population based on those counts. In using these formulae it must be assumed that the population is closed; that is, all animals that enter (by birth or immigration) or leave (by death or emigration) must be accounted for or else the population estimate will be increased or decreased, respectively. Park biologists partially controlled for these factors by omitting young of the year because this age group experienced some mortality between the two counts (Olympic National Park 1984:16). They did not, apparently, account for the possibility that animals might have immigrated or emigrated.

The first step of the population estimation procedure (accounting for less than 100% sightability) is apparently as reasonable as can be expected. What of the second step: sampling the park and calculating population estimates? The blocks examined in the sampling strata were chosen "randomly" (Houston et al. 1986:131–32), which means that each block had the same chance of being chosen for inspection as every other block within a sampling stratum. Six of the twenty-six total blocks in the low-density stratum were chosen; this number "represented a guess at the minimum [number of blocks] required to calculate an acceptable estimate" (Houston et al. 1986:132). So 23% of the blocks of the low-density stratum were chosen, representing 21.7% of the total area of the stratum. Twenty-eight of the fifty-nine total blocks in the medium-density stratum were chosen because twenty-eight blocks was "the maximum that was economically feasible to count" (Houston et al. 1986:133). So 47.5% of the blocks of the medium-density stratum were chosen, representing 47.7% of the total area of the stratum. This appears to be a reasonable sampling procedure, except that it is unlikely that a single low-density block that actually had a high density of goats (if such a block existed) would be chosen because so few of the blocks in the low-density stratum were examined. Discovery of such a high-density block would suggest that the strata were poorly defined, which would seriously lessen the accuracy of the estimate.

Problems with Procedures

The reason to define subpopulations of blocks is to control the heterogeneity in a population analytically. Stratifying a population "tends to improve population estimates, provided that the variable used for stratification is related to the subject of study" (Dixon and Leach 1978:16). Stratifying on the basis of the (suspected) density of goats is reasonable to obtain an estimate of the total population. Stratifying a population typically reduces the sample size needed *if the strata are defined in such a manner that the heterogeneity of the population is controlled by creating internally homogeneous strata or subpopulations.* Ideally, variation within each stratum will be minimized by the process of stratification, whereas variation between strata will be maximized. The park biologists discovered that at least the medium-density stratum was poorly defined: the "observed goat densities in two blocks of the putative medium density stratum greatly exceeded those in the known high density stratum. These [were] rude surprises" (Olympic National Park 1984:9).

The published version simply notes that "observed densities in two blocks exceeded those in the known high density stratum" (Houston et al. 1986:135). Both versions also mention that a rather different estimate is obtained if the sampling strata are made more homogeneous. Including the two surprising medium-density blocks in the high-density stratum and recalculating all values produces a population estimate of 1,036 goats (Houston et al. 1986:135; Olympic National Park 1984:9). Not only are 139 fewer goats estimated, but the error is lowered from ± 171 goats for the published 1,175 estimate to an error of ± 111 goats for the 1,036 estimate. Nonetheless, the larger estimate is the one most often mentioned by park personnel.

The lower (and seldom publicized) estimate of 1,036 goats results in part because removing the two surprising blocks from the medium-density stratum leads to a reduction of the average density of goats in that stratum. Including all twenty-eight blocks surveyed in the medium-density stratum, the average density is 0.02983 goats per hectare. With the two surprising blocks omitted, the average density is 0.02091. Thus, omission of those two blocks reduces the average density by 30%. The estimate for the medium-density stratum drops from 882 goats when all twenty-eight blocks and 29,860 hectares are included to 603 goats when twenty-six blocks and 28,848 hectares are included.

These numbers can be disputed because the values are inaccurate to some unknown degree. In calculating the average goat density per hectare I had to assume each block was 506 hectares in size because the park biologists simply indicated that the average size of all of the blocks was 506 hectares, but some blocks were larger and some were smaller (Olympic National Park 1984:6).

The unpublished document contains a map of the sampled blocks with frequencies of goats observed in each plotted. While the park biologists examined twenty-eight medium-density blocks, only twenty-six are indicated on their map (Olympic National Park 1984:8). In the published version of the same map (Houston et al. 1986:132) twenty-seven of the twenty-eight examined medium-density blocks are shown. Thus I had to determine the number of goats for the twenty-eighth block by calculating backward from the known values in the equations used by the biolo-

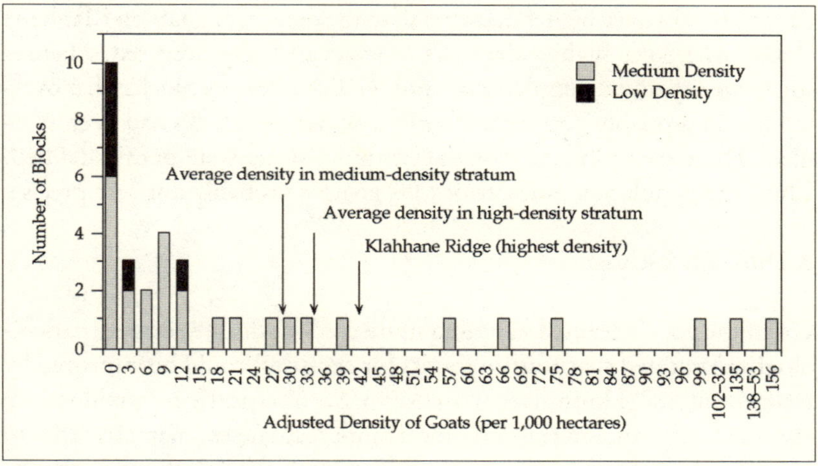

Figure 3.2. Frequency distribution of density of goats per 1,000 hectares in the sampling blocks of the medium-density sampling stratum and in the sampling blocks of the low-density sampling stratum (data from Olympic National Park 1984 and Houston et al. 1986). Note that the majority of the sampled blocks in the medium-density stratum have lower than average densities, that the low- and medium-density strata overlap, and that seven medium-density blocks have higher densities than the average for the high-density stratum.

gists. The block not indicated on the published map contained (by my calculations) an estimated three goats.

There is yet another issue regarding the procedure that warrants comment. David Davis and Ray Winstead (1980:245) caution that using the density of organisms per sampled area to estimate the total population demands the assumption that habitat variation is insignificant. The density of goats across the medium-density blocks that were examined varies tremendously. This variability is shown in figure 3.2, which shows the number of goats per 1,000 hectares and the number of blocks where each population density category is found (I used 1,000 hectares simply to omit decimal points). In calculating the density per block figures plotted in figure 3.2, I had to assume that each block was 506 hectares in size, as noted earlier.

Clearly, the majority (19 of 28 or nearly 68%) of the medium-density blocks fall well below the average density for that stratum of 29.8 goats per 1,000 hectares. Seven (25%) of the twenty-eight sampled medium-density blocks have densities of goats that are higher than the average density in the high-density stratum. And six, not two, of these supposed medium-density blocks have higher densities than the density of goats on Klahhane Ridge, where the highest densities of goats allegedly occurred. There is some overlap of the sampled low- and medium-density blocks; that overlap would probably be greater if more low-density blocks had been sampled. These data indicate that the sampling strata were poorly defined. Thus, the population estimate of 1,175 goats is probably not very precise.

A Different Estimate

Given the poorly defined sampling strata used for the 1983 census, park biologists indicated that "future counts [or censuses] could be improved by restratification" (Houston et al. 1986:135). Such restratifying could reduce the standard error and thus provide a more accurate estimate. In order to ascertain the potential magnitude of the error in the 1983 population estimate, I performed a new analysis of the data collected during the 1983 census for the suspected low- and medium-density sampling blocks.

First, I examined the frequency of both low-density and medium-density blocks that produced particular observed counts of goats (see fig. 3.2). Because the greatest number of goats in a low-density block as defined by park biologists was four, I chose that frequency as the division point between low- and medium-density blocks in order to restratify the sample of thirty-four examined blocks. Thus, all blocks in which four or fewer goats were observed were considered in my analysis to be low-density blocks ($n = 22$), and all blocks in which more than four goats were observed were considered to be medium-density blocks ($n = 10$). In addition, the two blocks in which the highest frequencies of goats were observed (with forty-five and fifty-two goats) were placed in a high-density stratum for my analysis. Such stratification after examination of the sampling blocks must be done with caution because the blocks examined may not be representative of the population and thus, again, the strata may be defined in a manner that does not reflect reality (Dixon and Leach 1978:21).

Once the strata were redefined, I calculated the proportion of the thirty-four examined sampling blocks that belonged to each sampling stratum (table 3.2, column B). I multiplied the total number of eighty-five (examined and not examined) sample blocks by the proportional frequency of examined blocks per stratum to determine how many blocks might be expected in each of my sampling strata (table 3.2, column C). The analysis may be weak here, as there is no way to assess the accuracy of

TABLE 3.2.
Population Estimates in Restratified Sampling Blocks

	A Number of blocks examined	B % of Col. A	C Col. B × 85	D Number of blocks not examined	E Col. D × 506	F Avg. density × Col. E
low density	22	65	55	33	16,698	70
medium density	10	29	25	15	7,590	354
high density	2	6	5	3	1,518	221

the underlying assumption that the proportion of examined blocks in each sampling stratum is equivalent to its proportion in the total population of eighty-five blocks. I subtracted the number of examined blocks per stratum from the total number of blocks per stratum to determine the number of unexamined blocks per stratum (table 3.2, column D). I multiplied the number of unexamined blocks per stratum by 506 hectares per block (column E) and then multiplied the resulting value by the average density of goats per hectare (adjusted for 0.66 sightability) per sampling stratum as determined by assuming each block was 506 hectares in size (table 3.2, column F). Summing those values gives a total of 645 goats for the park biologists' low- and medium-density blocks or my low-, medium-, and high-density blocks. My population estimate of 645 goats is 272 less than their estimate of 917.

Presuming the total counts and Klahhane Ridge population figures provided by the park biologists are accurate, adding them to my total of 645 brings the estimate of the total population to 903 goats. This figure is remarkably similar to the high end of the population estimate of 550 to 880 individuals provided by Stevens (1980) three years earlier, unlike the park biologists' estimate of 1,175 goats. Further, using my estimate of 903 goats and subtracting the 326 goats that were removed between 1984 and 1989 leaves only 188 goats unaccounted for in 1990 rather than the park's 460 (see next section). These observations suggest that my strata may be more accurate than those constructed by park biologists.

The park biologists indicate that their 1983 population estimate of 1,175 goats has a standard error of ± 171. A standard error cannot be calculated for the population estimate I have derived because relevant data such as the precise size of the various sampling blocks that were examined are not provided in the documents I have examined (Houston et al. 1986;

Olympic National Park 1984). Thus I cannot argue that my estimate is more accurate. What is important here is that in both cases these are *estimates.* It will become clear that they are not treated as such by the park biologists and spokespeople.

As noted above, the sampling blocks were chosen randomly. In terms of procedure, those blocks should have been examined in the order in which they were chosen. In many situations where random sampling is used, examination of units in this order is not practical. It could, however, be critical in estimating the size of the mountain-goat population in Olympic National Park. All census activities in 1983 took place from a helicopter. In the unpublished version of that census, park biologists noted that the goats were disturbed by the noise of the helicopter and sometimes hid in vegetation while it was still some distance away (Olympic National Park 1984:16). In the published version they assume that this problem was insignificant: "Observations of goat behavior, plus the continued successful trapping suggest that the effects of local helicopter traffic were minimal" (Houston et al. 1986:133). This contradicts more detailed findings (Coté 1996), but helps maintain the appearance of estimate accuracy.

Hiding goats would reduce the size of the estimate. Movements of goats from one sampling block to another could increase the size of the estimate because individual goats might be counted more than once (which would violate various statistical assumptions). Such significant movements in response to helicopter surveys have been documented for mountain goats in other areas (Coté 1996). While it is clear that the mountain goats in the Olympics move great distances (Stevens 1983), it is unclear if they traverse significant distances due to helicopter disturbance. However, they may not have to move very far to influence the Olympic population estimate. Of the thirty-three total medium- and low-density blocks examined by the biologists,[2] only eleven (33%) were geographically isolated from other examined blocks. Twenty-two sampling blocks were adjacent to at least one other sampling block; thus, a goat could have been counted in one block and then moved to another block, where it was counted again. There is unfortunately no way to evaluate the potential effects of this factor.

---

[2] Note that while thirty-four blocks were examined, the location of one of these is not indicated in the documents I have examined. Therefore, only the locations of thirty-three blocks relative to one another can be considered here.

POPULATION REDUCTION

After collection of information for the 1983 population estimate, 326 goats were either captured and relocated by park biologists, died or were euthanized as a result of capture-related activities, or were legally or illegally killed (Houston, Schreiner, Moorhead, and Olson 1991:90). Methods identical to those used to generate the 1983 population estimate were used in 1990 to estimate the population size. This time park biologists estimated the population at 389 individuals. Park biologists indicate that "known goat removals totaled 326 animals from autumn 1983 through autumn 1989" (Houston, Moorhead, and Olson 1991:215; Houston, Schreiner, Moorhead, and Olson 1991:90): 245 were removed by the park service, 80 were legally harvested, and 1 was illegally killed. Houston, Schreiner, Moorhead, and Olson (1991) mention the 81 goats that were removed by people other than park-service biologists in the text, although they are not listed in the table. But their table lists a total of 407 goats as having been removed between 1981 and 1989, inclusively. Perhaps that is why some of the confusion exists.

After the 1990 population estimate was publicized, Jon Luoma wondered:

> How could removing just 400 animals bring down the population [of 1,175] by 800? Part of the answer may be attributed to normal population fluctuations or to the difficulty of obtaining an accurate count due to the goats' habit of moving in and out of dense, brushy areas below the treeline, where they're hard to see. But part of the answer, too, could be that many offspring died, at least indirectly, when adult females were captured. (Luoma 1993:37)

Luoma underscores the fact that published statements of population size involve estimates of questionable accuracy. When he says that 400 goats were removed I suspect he has made the mistake of quoting the total number of 407 goats removed between 1981 and 1989 inclusively (Houston, Schreiner, Moorhead, and Olson 1991:90). The park biologists indicate that the number that should be used is 326 (Houston, Schreiner, Moorhead, and Olson 1991:215), so Luoma's question is even more pressing. The correct math leaves 460 goats unaccounted for rather than the incorrect 379 if Luoma's figures are used.

Even the park biologists wondered about the missing mountain goats. They believe that 67 of them were kids that were still nursing when their mothers were captured and "suspect few of [these] kids, if any, survived" (Houston, Moorhead, and Olson 1991:215). That brings the number unaccounted for down to 393 (= 460 – 67). These 393 goats, the park biologists "suspect," were the victims of "substantial winter mortality [because] winter snowpack returned to more 'normal' levels, following the unusually low levels from about 1976 to 1981" (Houston, Moorhead, and Olson 1991:215). Such an impact on mountain-goat populations has been documented elsewhere.

Park biologists directly removed or can account for 326 goats (27.7% of the estimated 1,175 goats present); they also suspect that the removal of lactating nannies accounts for the loss of 67 (5.7%); and finally, harsh winters account for a loss of about 393 goats (33.4%).

If, as the park biologists suggest, winter mortality is density independent (that is, "normal snowpacks" result in the death of goats regardless of their numbers or density on the landscape), a 33% decrease in the population of goats will occur every six years when the winter snowpack is "normal." Thus, given sufficient consecutive years of normal winters, the goats will be eradicated from the park without human intervention (see below). Did the goat-removal program continue into the late 1980s because we lack the ability to predict whether future winters will be more harsh or less harsh than "normal"?

MINIMUM VIABLE-POPULATION SIZE

Park managers desire the smallest possible (if any) mountain-goat population in the park. State Department of Wildlife personnel want a goat population that could sustain annual harvesting of eight to ten individuals. This is a classic problem in wildlife management and requires research to determine the minimum viable-population size that could also withstand some hunting pressure. "Minimum viable-population size" is a slippery term: there is no clearly agreed upon measurement method (Nunney and Elam 1994; Soulé 1987b). The term often seems to mean the smallest possible population size that will survive typical, natural, life-threatening forces over some multigenerational period, particularly when sufficient time for potential changes in environment is included (Nunney and Elam 1994).

But simply surviving as a population is probably an insufficient criterion, and some biologists argue that the maintenance of a certain level of genetic diversity within the population is equally important or more important than simple numbers of individuals (e.g., LaCava and Hughes 1984; Soulé 1987a).

Although population size and genetic diversity are closely correlated (Frankham 1996), there are biological reasons to adopt the position that genetic diversity is more important (Lacy 1997) while realizing that it is populations of individual organisms that are managed. For example, the genetic diversity of a population often decreases as the size of the population decreases (Lande 1988); the causes of population decrease can be many and varied and include higher rates of loss (by natural mortality, overhunting, or removal of individuals for transplanting) than of recruitment (the addition of new individuals from birth or immigration or both). After population decrease—a so-called bottleneck—if the population increases, rates of inbreeding increase and genetic variation may or may not increase (Lacy 1997; Lande 1988); the probability that the population will survive random fluctuations in environmental conditions may decrease because decreased genetic diversity has been shown to result in increased mortality of young, reduced fertility in adults, and a loss of potential for at least some individuals in a population to be (pre)adapted to future environmental conditions (Samson et al. 1985; Soulé 1987a). Such bottleneck effects have been discussed with respect to mountain goats (Naylor et al. 1990; Smith and Raedeke 1982).

Typically, the smaller a population, the less its potential to have a large range of genetic diversity (Frankham 1996; Lacy 1997; Lande 1988). While the converse is not necessarily true, it often is true given sufficient time for the population to produce offspring with new genetic mutations and for nonlocal individuals originating from other populations with a different gene pool to interbreed with the subject population. Of course, this option is unavailable if the subject population is geographically and thus genetically isolated from other populations of the same species (e.g., Bleich et al. 1990; Naylor et al. 1990). Reducing the size of a subject population thus has two effects on its genetic diversity: that diversity will be reduced simply because there is less genetic material in the population and because fewer individuals mean fewer chances for mutations and fewer chances for interbreeding with nonlocal individuals.

Russell Lande (1988) argues that while genetic variability is an important consideration when determining a minimum viable-population size for a species, that variability may be less important for survival of the population than simple numbers of organisms or demographic parameters (Berger 1990). The genetic bottleneck and the depression in genetic diversity caused by inbreeding may be irrelevant to extinction events if a population becomes extinct for "demographic reasons" (Lande 1988:1457), including modified social interactions between individuals, modified survival probabilities of various age-sex categories, and modified home-range characteristics (see also Berger 1990).

The research effort that must be made to derive such values (Korn 1994) is only the first hurdle; as Michael Soulé (1987a:5) indicates, "certain groups in society might be content with a 50% probability of persistence [of a species] for 100 years, while other groups would settle for nothing less than a 99% probability of persistence for 1000 years." If one is going to remove individuals from a population, one needs to have estimates of the population size in order to ensure that one does not remove so many individuals that the remaining population will be too small to be able to reproduce and survive. This, of course, presumes that one knows the minimum viable-population size of the species. Has the park carried out the necessary research?

One management alternative described in Olympic National Park's 1987 goat-management report involves removing all goats from the core of the park but retaining a population in Olympic National Forest along the eastern periphery of the park, where goats can be hunted. Park biologists note that this management alternative requires the maintenance of a "huntable population" (Olympic National Park 1987a:52–53, 59). "The major issue would probably center around the determination of [eastern park] boundary subpopulation levels that would satisfy both NPS and Washington Department of Wildlife requirements. Setting [population] levels that would simultaneously reduce environmental impacts in the park and maintain a huntable population outside the park will require a cooperative research effort" (Olympic National Park 1987a:60). These statements suggest that park biologists have no idea what the minimum viable population should be. So far as I am aware, no research on what the minimum necessary population size might be has been undertaken by any of the agencies involved: Olympic National Park, Olympic National Forest, and Washington Department of Wildlife. How can the park justify reducing the Olympic mountain-goat population so drastically ([1,175 – 389] ÷ 1,175 = 67%)?

A monograph on Olympic's mountain goats prepared by park biologists (Houston, Schreiner, and Moorhead 1994) contains no mention of minimum viable-population sizes. The authors estimate that if a huntable and viable population is maintained on the eastern side of the peninsula, 250, 200, or 150 to 200 goats (Houston, Schreiner, and Moorhead 1994:199, 203, 204) might remain in Olympic National Forest. These estimates, whose minimum and maximum values differ by 66%, indicate that the biologists do not have a very good idea of how many goats would represent a minimum viable population.

The Draft Environmental Impact Statement (Draft EIS) prepared by park personnel (Olympic National Park 1995) is not much better. It states that "management action within the park could have implications for recreational hunting of goats on other peninsular lands" (Olympic National Park 1995:8). If the entire population is reduced by 80%, from the original *estimated* high of 1,100–1,200 to 200, I have little doubt that park actions will influence recreational hunting outside the park. The Draft EIS (Olympic National Park 1995:50) also states that "the goat population outside the park may or may not remain viable long-term under" this management alternative. This statement alone indicates that the park biologists do not know the size of a minimum viable population. In fact, park management actions may already have doomed the Olympic goat population.

IMPLICATIONS OF POPULATION REDUCTION

The estimated 1983 population of 1,175 individuals was reduced by 1990 to an estimated 389 individuals. I have mapped the changes in range occupied between 1983 and 1989 in figure 3.3. The actual 1990 population might have numbered as few as 181 or as many as 597 individuals (*IGMT Newsletter* 2:2). As noted above, 326 goats are known to have been removed, and park biologists suspected that another 67 nursing kids died when their mothers were removed. Thus, they could account for the removal of only 393 individuals from the population of 1,175. This left an additional 393 goats that the biologists could not directly account for. They suggested that those individuals may have been the victims of "winter mortality" because winter snowpack "returned to more 'normal' levels" after 1981.[3] Therefore, by the reasoning of the park biologists, 33% of the goat population was removed by "normal winter snowpack" over six years.

---

[3] Of course, this means that the increase in the population documented between 1976 and 1981, when winter snowpack was less than "normal," could be attributed to mild winters and in turn that the estimated maximum population of 1,175 individual goats was hardly "normal" and would have subsequently decreased significantly without human intervention.

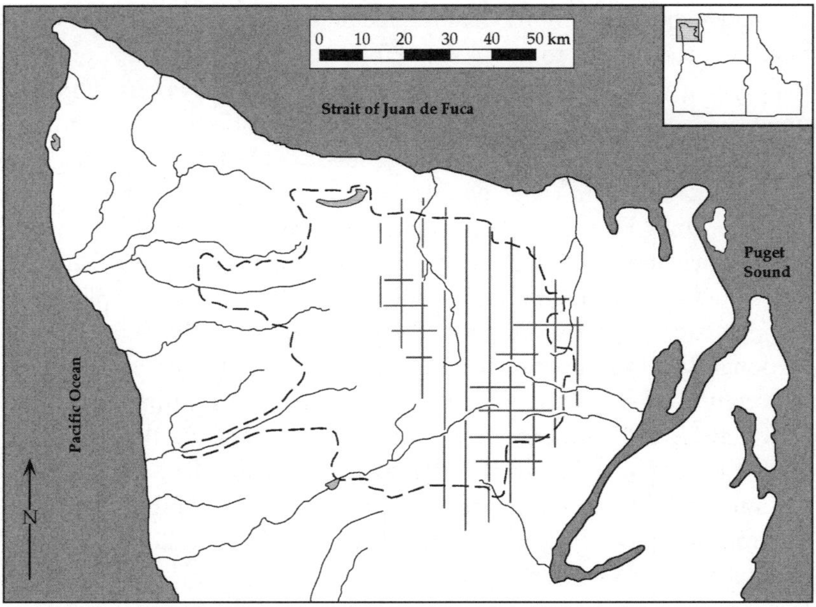

Figure 3.3. Comparison of the 1983 range (vertical hatching) and 1990 range (horizontal hatching) of mountain goats on the Olympic Peninsula (based on data in Moorhead and Stevens 1982 and Houston, Moorhead, and Olson 1991). The dotted line approximates the boundary of Olympic National Park.

Is a population of 389 mountain goats enough to maintain a minimum viable-population size for the eastern peninsula? Park biologists fully intended to continue to remove goats in 1990, 1991, and 1992 (Houston, Schreiner, Moorhead, and Olson 1991) while simultaneously maintaining a huntable population. I use the lower estimate of the 1990 population size (181 individuals) in the following discussion, to account in part for the fact that the park managers intended to reduce the population further. And I assume that by "normal" winter snowpack the park biologists mean an "average" winter snowpack. If that assumption is correct, then it is easy to show that the remaining mountain-goat population may well be doomed regardless of the activities of park biologists.

Park biologists assume that mortality is density independent; if the winter snowpack was normal between 1990 and 1996, then 33% or 60 of the remaining 181 mountain goats would have been eliminated, leaving 121 individuals in 1996. If winter snowpack continues to be normal, another 33% will be removed between 1996 and 2002, leaving 81 goats. If

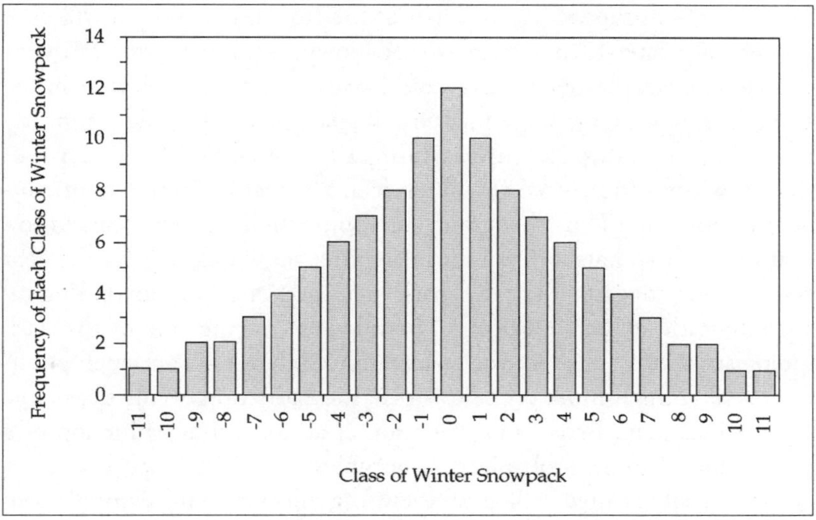

Figure 3.4. A modeled normal distribution of twenty-three classes of 110 fictitious winter snowpacks. Snowpack classes designated by negative numbers indicate that snowpack is less than average; snowpack kinds designated by positive numbers indicate that snowpack is greater than average.

the pattern continues as long as the winters are normal (or worse), by the year 2032 the size of the mountain goat population will be back to its starting point of 11 or 12 individuals. Of course, winters may be less harsh or more harsh than normal. But because a statistical average is typically in or near the middle of a frequency distribution, then *more than half* of the cases in that frequency distribution will be sufficiently harsh to affect the goat population adversely. This can be shown by example.

I plotted the frequency of each of twenty-two classes of winter snowpack over 110 years in figure 3.4. This is a fictitious example to illustrate the points made in the preceding paragraph. I plotted equal numbers of harsh winters and mild winters in such a manner as to produce a graph that approximates what statisticians call a normal or bell-shaped curve. The horizontal axis of the graph designates the class of winter snowpack, with 0 denoting a normal or statistically average snowpack, classes with minus signs denoting less-than-average snowpack, and classes with positive signs denoting greater-than-average snowpack. Note that most of the winters closely approximate the average winter snowpack, few winters are extremely mild or extremely harsh, and an average winter snowpack is the most frequently represented class.

This graph shows that over half of the 110 winters plotted will have average or greater-than-average winter snowpacks. In this graph 61 winters (55% of the plotted winters) could result in a natural decrease in the size of the mountain-goat population. If goats are lost by mild but nearly normal winter snowpacks such as during the class labeled "–1," then 71 of the 110 winters (nearly 65%) will result in a reduction in the mountain-goat population. Thus this graph, in conjunction with the conclusion that normal and harsher than normal winter snowpack reduces the goat population, suggests that the park managers are playing ecological roulette with an unfair wheel.[4] The odds are on the side of the park biologists, who wish to see the goats removed. Not only are small populations of animals more prone to stochastic extinction events than large populations (e.g., Berger 1990; Samson et al. 1985), but extinction of a population is more probable if it occupies a limited range because a geographically limited ecological event can affect the entire population.

In all fairness to the park biologists, there are many reasons why this may be an inaccurate characterization of the potential future of the Olympic mountain goats. The contraception and sterilization panel convened in October 1991, for example, speculated that the population might increase significantly during the interim between the cessation of the goat capture and transplant program in 1989 and the initiation of new population-control techniques several years later (Warren et al. 1992). If winter mortality due to snowpack is dependent on the size of the population, then as the population decreases in size the probability that at least some goats will survive the winter will increase and the proportion of the population that dies will decrease. In the Eagle Cap Wilderness Area of northeastern Oregon (Vaughan 1975), for example, competition for limited winter range is reduced when fewer goats compete with one another for the resources found in that range. Further, the Eagle Cap population was started with a total of only seven transplanted individuals in 1950. The total descendant population has apparently never numbered more than about three dozen individuals (five goats were harvested each year for four years during the mid-1960s), yet it seems to be holding on despite the clearly limited genetic diversity such a small population could contain (Vaughan 1975).

---

[4] It was park biologist Ed Schreiner (quoted in Burger 1987:28) who suggested that transplanting "exotic" mountain goats onto the Olympic Peninsula amounted to "playing ecological roulette." I find the metaphor appropriate in a different context.

The problem is that we do not yet know precisely what limiting factors influence mountain-goat survivorship and mortality in the Olympic Mountains, though some evidence suggests that winter forage "*may*" be a significant factor (Houston and Stevens 1988:235; emphasis in original). Clearly, more research must be done on such factors.

Research undertaken during the 1980s shows that many species of animals have been lost from national parks in the United States and Canada where plant and animal species are supposed to be protected from extirpation (e.g., Glenn and Nudds 1989; Newmark 1986a, 1986b, 1987, 1995). We have insufficient information about many of our national parks to determine the kinds of organisms that might have been lost (or gained) over the last century or so (e.g., Stohlgren and Quinn 1992). It is clear that the deficiencies of the historical record can often be filled via paleontological research (see Olson 1989 and Steadman 1995 for recent reviews of insular faunas in particular). As the fossil record becomes better known, it regularly indicates errors in biogeographic histories written on the basis of twentieth-century data alone. It also indicates that extinction was a very regular event among insular faunas.

The estimate for the 1994 goat population was about 288 individuals (Houston, Olson, Hoffman, and Moorhead 1994; see below), perhaps 100 less than in 1990. If snowfall during the winters of 1990–91, 1991–92, 1992–93, and 1993–94 was greater than "normal," this would be possible proof of the effects of such snowfall. Park biologists, however, argue that the 1994 population is not statistically different from the 1990 population (see below) and is so low because goats were in heavy vegetation to escape higher-than-normal summer temperatures and thus were not as visible as in other census years. Thus there is no apparent reason to test the effect of above-normal winter snowfall. The possibility of testing by comparison with temperatures during the 1983 and 1990 censuses is not discussed (Houston, Olson, Hoffman, and Moorhead 1994). There seems to be no rigorous research design involving the testing of hypotheses, the structured generation of data with which to test hypotheses, and hypothesis refinement and additional testing—a conclusion also reached by others (Peterson et al. 1994).

ESTIMATE ACCURACY

The population estimates described above were discussed in the *IGMT Newsletter* 2 (June 1992:2), circulated to 600–700 people. The discussion was a response "to some questions most frequently asked at public meetings." One question and its answer are phrased as follows in the newsletter:

> How reliable are the population estimates of 1175 ± 171 mountain goats on the Olympic Peninsula in 1983, and 389 ± 106 in 1990?
>
> —These counts are estimates only, and are reliable only to show broad trends. The estimates were from sample counts of goats in portions of the Olympic Peninsula, which were then extrapolated to yield a population estimate for the entire goat range. Using [statistical techniques described above tells] us that with 95% certainty the true population size in 1983 was between 840–1510 goats, and in 1990, between 181–597. These wide "confidence limits" show that the actual numerical estimates of population size cannot be interpreted literally.

This is not the way the estimates have been used and interpreted by park personnel, however. Janis Burger (1987:28), a park employee, states explicitly in an article in a popular magazine that "the goat population at the park and the adjoining Olympic National Forest now numbers almost 1,200." She includes no statements about the probability of the correctness of this estimate, about population trends, or about the probable range of actual population size. Burger is interpreting the 1983 population estimate literally. Even the park biologists seem to believe that the estimates of 1,175 and 389 are somehow real and accurate. Why else would they so diligently attempt to account for the difference between the 326 goats they removed from the population of 1,175 (and the 67 kids that died as a result) and the remaining 393 goats that they cannot directly account for?

Statistically, there is a 95% chance that the 1983 population consisted of only 840 individuals, which is no different from Stevens's (1980) maximum estimate that the population numbered 884 only three years earlier. Why do park biologists conclude that "the census indicates that the goat-management problem facing the National Park Service is very likely larger than anticipated initially" (Houston et al. 1986:136)? I suspect that they are interpreting the population estimates literally. Certainly they are not monitoring "broad trends," except to note that the 1990 estimate of 389 goats is much smaller than the estimate of 1,175 for 1983 (Houston, Schreiner, Moorhead, and Olson 1991). But we could have known prior to the census of 1990 that the population would be smaller than in 1983 simply because we knew that 326 adult goats and 67 nursing kids had been removed.

The *IGMT Newsletter* states that "estimating the size of mountain goat populations and other wildlife species has consistently been a difficult problem for wildlife managers throughout North America. The mountain goats of the Olympic Peninsula are just one example of this." If estimation is difficult, then of what use are the estimates? Apparently, they were made because "park managers were considering an extensive goat removal program as one management option, [therefore] a more comprehensive estimate was needed" (Houston et al. 1986:131). Such an estimate, if accurate, would indicate the magnitude of the management actions necessary.

A final indication that the wildlife biologists studying the mountain goats view the population estimates as accurate is found in the unpublished report of a more recent aerial census in July 1994 (Houston, Olson, Hoffman, and Moorhead 1994). Procedures were similar to those followed in the 1983 and 1990 censuses, with the important difference that the sampling blocks were restratified to reflect the most recent knowledge about goat distributions. Census activities produced an estimate of 288 ± 63 goats. The park biologists statistically compared this estimate with the 1990 estimate (389 ± 106 individuals) and concluded that the two estimates do "not differ significantly [statistically]" (Houston, Olson, Hoffman, and Moorhead 1994:6). This is because of the large errors (the ± values) associated with the estimates.

In making such a statistical comparison of the two estimates, park biologists are implicitly assuming that the estimates are real. This is also evident when the park biologists note that the 1994 census was undertaken "during a period that included record high temperatures for the Puget Sound region" and that "goats are known to be sensitive to high temperatures and may seek relief in forests" (Houston, Olson, Hoffman, and Moorhead 1994:6). They are implying that the 1994 estimate is probably too low to be trustworthy because of decreased visibility of goats and that the apparent loss of 111 goats (389 – 288), or 28.5% of the 1990 population, over the succeeding four years is a function not of any real population trend but of census timing. If the biologists did not conceive of these estimates as real, how could they conclude that the 1994 estimate is inaccurate? This in turn calls into question their conclusion that the 1990 and 1994 censuses produced results that are not statistically different.

Two values are essentially equal only if both are equally accurate; if they are not, then one cannot know if the two values are truly equal or not.

Park researchers again estimated the size of the goat population in July of 1997. Their estimate, thus far found only in an NPS news release issued on July 31, 1997, is that "with 95% certainty, the current population [is] between 237 and 325. This estimate is statistically no different than either the 1994 census ... or the 1990 estimate." Interestingly, this news release suggests that the 1,175 goats estimated to occur in the park in 1983 may have resulted from the "unusually low snow years ... between 1976 and 1981."

A FINAL ESTIMATE

The statement that the present mountain-goat population "would likely increase to approximately 1400 animals" if it were left alone is repeated at least five times in the Draft EIS (Olympic National Park 1995), but it is not justified or documented in any rigorous fashion, although the authors note that this estimate concerns "the maximum sustainable population for goat habitat *in the park*" (Olympic National Park 1995:24; emphasis added). The monograph produced by the NPS also indicates that the goat population in the park could reach as high as 1,400 (Houston, Hoffman, Olson, and Moorhead 1994:199). Both works cite the same unpublished report. The monograph goes on to note that "because about 15% of the goat habitat is outside the park [the total population for the Olympic Mountains could be] 1650."

These estimates of 1,400 to 1,650 goats prompt two comments. First, why were these estimates not mentioned before? Why are they included only in the two documents that the park considers critical to resolution of the controversy? The second comment concerns the estimate of 1,400 goats. Hoffman and Wright (1987) describe the procedure they used for deriving this estimate:

> the [total goat] population was divided among 21 separate [geographic] units which have either biological (some are essentially discrete populations of animals) or management significance (zone boundaries are based on political or logistical constraints) [see fig. 1 in Houston, Schreiner, Moorhead, and Olson 1991 for a map of these units]. These encompass most of the important goat habitat on the Olympic Peninsula.... Within each unit, the numbers of mountain goats of each sex in each age class were

> modeled. [That is,] a best estimate of population for each area [or zone was] determined through a process of combining information from aerial counts, ground surveys and educated guesses by the authors and other Olympic National Park researchers. (Hoffman and Wright 1987:47, 50)

Hoffman and Wright's (1987:51) table 13 indicates that they determined the area of each zone or spatial unit, estimated the *maximum* number of individual goats that *could* live in each area, and summed the number of goats estimated for each zone to obtain their estimate. The second step was accomplished by using information from "aerial surveys, ground counts and educated guesses" (Hoffman and Wright 1987:50).

As we have seen, park researchers were in many cases badly mistaken about how many goats were found in particular areas. Therefore, I have little faith in the estimate of 1,400 animals for park lands and even less faith in the concomitant estimate of 1,650 for the entire Olympic Peninsula. According to park biologists, if 15% of the habitable goat area is outside the park boundaries, then 15% of the goats (an estimated 247 individuals) should be there. To derive this estimate, park biologists have assumed that the density of goats is uniform across park and nonpark lands. Their own data indicate this is an inaccurate assumption. But the possibility that there may eventually be 1,650 goats "indicates that the goat management problem facing the NPS [could be] larger than anticipated initially" (Houston et al. 1986:136). Such a possibility is certainly more worthy of action than a mere 1,175 goats.

4
PLANTS

The reports prepared by Olympic National Park biologists typically contain statements such as, "plant communities in the Olympic Mountains have evolved without goats" (Olympic National Park 1981:8); the Olympic alpine ecosystem has "evolved over thousands of years" (Olympic National Park 1981:5);[1] and, "As goats increase [in number], disturbance to vegetation and soil from their foraging and trampling is also increasingly evident" (Olympic National Park 1981:5). Such observations are made relevant to the controversy by noting that "prevention and control of damaging activities by man to natural ecosystems has evolved as a cornerstone of National Park Service policies" (Olympic National Park 1981:7). Park personnel argue that humans are responsible for the presence of mountain goats in the Olympic Mountains. Where is the evidence that the Olympic alpine ecosystem has been disturbed? Until recently, it has resided almost totally in unpublished park-service reports. The published exceptions are remarkably few in number (Houston, Schreiner, and Moorhead 1994; Olmsted 1977, 1979; Pfitsch and Bliss 1985; and perhaps Houston and Stevens 1988).

Two effects on the Olympic alpine ecosystem are commonly attributed to mountain goats: lowered plant biomass from grazing and trampling and exacerbated erosion from wallowing (discussed in chapter 5). The plant data appear to have served as the first documentation that the goats were at the base of a perceived ecological crisis.

---

[1] Whether ecosystems in fact "evolve" is debated (e.g., compare Bambach and Bennington 1996 with Stenseth and Smith 1984). One position concludes that ecosystems and biological communities do not evolve; they merely change composition. The other position holds that communities evolve in concert with changes in species of organisms (e.g., Brett et al. 1996). Given the debate, I allow park biologists their use of the term "evolve," whereas I use the term "develop" or "change."

MOUNTAIN GOATS AND PLANTS

Victoria Stevens (1983:8–9) indicates that the rutting season of the Olympic mountain goats begins at the end of November and extends into early December. Gestation is about six months, and most births occur during the last two weeks of May (Stevens 1983:23). Lactation, which "lasts for approximately three weeks," demands that nursing nannies consume vegetation of high nutrient content; thus the nursing period tends to "coincide with the average time of greatest vegetative shoot growth" (Stevens 1983:9; see also Hutchins 1984). Given the high altitudes occupied by mountain goats, their utilization of the landscape and plants fluctuates with the seasons. The growing season of plants starts later in the year as one gains elevation. Vegetation on south-facing slopes grows earlier and matures before vegetation on north-facing slopes, which are covered with snow longer.

> Early in the summer, when female goats are lactating and require the greatest energy of any time of the year, the bands of females and young can be found feeding on high meadows of approximately south-facing slopes. These meadows are the first to melt from winter snow....Later in the summer, when south slopes are dry and hot, the north facing slopes provide the best protection from solar radiation and access to later-emerging vegetation. (Stevens 1983:28)

Several important variables must be considered when monitoring the effects of goats on alpine habitats. Minimally, we should be aware of the elevation of the habitat; the time of year when measurements on vegetation are taken; and the aspect (north-facing, south-facing, etc.) of the slope. Further, given that individual variations in each season tend to influence the amount of plant growth, we need to monitor or control these variables if we are to obtain accurate measures of the effects that grazing mountain goats have on plants.

We must know two other kinds of information as well. As W. A. Pfitsch et al. (1983:69) indicate, "The question arises as to what evidence there is that mountain goat activities have changed the plant communities of Klahhane Ridge. This is extremely difficult to answer in light of the fact that there was no detailed plant-community analysis done before the goats became abundant. Two areas [having small goat populations] were sampled with the intent of addressing this question." These two areas served as control samples. However, because of potential environmental differences between the nineteenth and the twentieth centuries and because the nineteenth-century plant community is unknown, their research is compromised from the start.

A further complication is that "there is a history of cattle grazing at Hurricane Ridge and sheep grazing at Deer Park" (Warren 1987), a fact not mentioned in any of the literature written by park personnel and generally available to the public. Grazing of domestic livestock on what are now national-forest and national-park lands began in the late nineteenth century; all early settlers of the Olympic Peninsula "had a few head of cattle," and some had sheep (Morgenroth 1991:47, 52). Leasing of grazing rights on forest lands prior to the establishment of the park in 1938 was called an "established success" early in the twentieth century (Morgenroth 1991:142). It is likely that the grazing and trampling activities of domestic livestock initiated the modification of the Olympic ecosystem, but the nature and extent of that modification are unclear. How is such a phenomenon to be dealt with in studies such as Pfitsch et al.'s (1983)? In the studies reviewed below, it is simply ignored.

EARLY STUDIES

The earliest study of the interaction of mountain goats and plants in the Olympic Mountains is by Ingrid Olmsted (1977, 1979). Her first publication is an abstract of a presentation given at the First International Mountain Goat Symposium (Olmsted 1977). The second of the four paragraphs in the abstract is reproduced here in full:

> Mountain goats, introduced in the late 1920's, have reached a population size that has affected the vegetation. The mountain goats have established themselves on Klahhane Ridge in greater number than anywhere else in the Park. Various types of influences from feeding to trampling to bedding and dust-bathing [wallowing] have been measured. Ratios of vegetated/unvegetated goat-used areas have been determined and will be compared to similar vegetation types that are little or not at all used by goats. Cover and density measurements were made of most plant species used by the mountain goats. Permanent plots have been established for monitoring changes and succession. (Olmsted 1977:200)

There are four key things to note here. First, goats in the park were at their densest concentration on Klahhane Ridge. Second, the vegetation was being affected in the mid-1970s because of the population size; recall that the 1973 estimate of the population indicated about 400 goats *in the entire park* (see fig. 3.1). Houston and Stevens (1988:230) indicate that the estimated population of goats on Klahhane Ridge during 1977 was 181. Third, the influences of grazing, trampling, and wallowing were *measured.* Fourth, comparisons between areas used by goats and areas not used by goats would be made.

In the first relatively detailed publication on the effects of mountain goats on the Olympic alpine ecosystem, Olmsted (1979:1143) indicates that goat "use pressure on Klahhane Ridge has resulted in some apparent disturbance of the vegetation." She also notes that the goat population on that ridge had "increased perhaps threefold during the last 15 years," that "the goats are more concentrated here than anywhere else in the Park," and that "studies of the vegetation prior to the introduction of the mountain goats were not made" (Olmsted 1979:1143). How can we know what impacts the goats have had on the Olympic alpine ecosystem when we do not know what that ecosystem was like prior to their introduction? Further, Klahhane Ridge cannot be used as a guide to the magnitude of the impact of goats in other areas of the park because it is the area with the highest density of goats. Goats may have been so densely crowded on Klahhane Ridge in 1977 in part because of human activities there (see chapter 5).

Olmsted (1979:1143) reports that "we estimated the whole [of Klahhane] ridge is inhabited by 180 mountain goats [and] we have about 21 mountain goats/km^2."[2] She indicates that "the erosion rate is accelerated by the goats" as they move across steep slopes (30 to 35 degrees) composed of slide rock and that "most of the feeding and traveling occurs on steep slopes, [therefore] loosening or removal of plants through trampling is greater than on flat terrain" (Olmsted 1979:1143, 1146). Olmsted observes that goats do use trails, but when humans approach too near, the goats move away, often rapidly, across steep slopes devoid of trails; this

[2] Since there are 100 hectares in a square kilometer, this is equivalent to 0.21 goats per hectare. Later estimates put the goat density at about 0.14 goats per hectare in a snow-free (May to November) range of 1,300 hectares (Pfitsch et al. 1983:55). Note that this population grew to about 220 goats by 1980; due to removal efforts by park biologists, this population consisted of 85 individuals at the end of 1983 and only 41 individuals in late 1986 (Houston and Stevens 1988). Only 3 goats were estimated to be there in July 1990 (Houston, Moorhead, and Olson 1991).

"hastened trampling" exacerbates erosional processes. She also describes the use, size, and frequency of wallows and salt licks (see chapter 5) and the bedding habits of the goats and ranks the frequency and amount of "natural" damage, from greatest to least, as trampling, dust-bathing (wallowing), bedding, and feeding (Olmsted 1979:1147).

Importantly, Olmsted (1979:1146) notes that "if we include the disturbance inflicted by the man-placed salt blocks, then the resulting destruction from the concentrated use [around salt blocks] would take first place." That damage is not natural and cannot be completely attributed to the behavior of the mountain goats, a fact noted by Olmsted but curiously absent from virtually all reports produced by park biologists. For example, such damage is said to be "minimal" in the Draft EIS (Olympic National Park 1995:22), but no evidence for this statement is provided there (see chapter 6).

Several early studies of human impacts were published (Bell and Bliss 1973; Moorhead and Schreiner 1979; Schreiner and Moorhead 1981), but several major studies written at this time were not (e.g., Schreiner 1974, 1982). I am unaware of any reports on the human influences on the park's mountain ecosystem completed after the early 1980s. All later efforts appear to have been directed toward studying the influences of mountain goats (e.g., Pfitsch et al. 1983; Pfitsch and Bliss 1985; Schreiner 1987).

Katherine Bell and Lawrence Bliss (1973) found that human trampling had significant impacts on the delicately balanced alpine plant communities of the Olympics. Both the amount of ground cover and plant biomass were significantly reduced by the equivalent of sixty people walking on vegetation over one month. Vegetation in intensively trampled areas recovered more slowly and less completely than vegetation in lightly trampled areas. Where roads had been cut into mountainsides, vegetation had not recovered in some thirty years. Bell and Bliss (1973:31) indicate that all this was due to "the severe [alpine] environment, short growing-season, relatively few [plant] species [present], [the] slow growth-rates [of the plants], and the general lack of successional [plant] species that in temperate regions rapidly seal the surface of plant-denuded sites." Importantly, they concluded that human use should be limited, "either by restricting the number of people or by providing asphalt paths to confine their walking" (Bell and Bliss 1973:32). Building more paths would surely exacerbate erosion and destroy plants. The *IGMT Newsletter* (2:3) indicates that both the park and the Olympic National Forest "are developing Wilderness Management Plans to minimize human impact on much of the land they manage."

Moorhead and Schreiner (1979) studied the impact of humans on backcountry campsites in the park. They found human trampling to be a very significant process that removed floral ground cover and recommended that campsites be restricted to locations with the greatest potential for floral recovery, that the number of minor trails be controlled by adding more trail markers and directional indicators, and that new campsites be eliminated. But were these recommendations followed? What are the results of other studies of human impact (e.g., Schreiner 1982)? Why have those results remained unpublished (with the single minor exception of Schreiner and Moorhead 1981)?

LATER STUDIES

Many studies of the influences of goats on the alpine flora of the Olympics after 1980 are unpublished (e.g., Pfitsch 1981; Pfitsch et al. 1983; Pike 1981; Reid 1983; Schreiner 1987). One exception (Pfitsch and Bliss 1985) contains some interesting observations and information from unpublished documents. The authors indicate, for example, that in 1980 the population of goats on Klahhane Ridge consisted of 180 individuals (0.138 goats per hectare), significantly less than the 0.21 goats per hectare indicated by Olmsted (1979). Why this discrepancy?

Houston and Stevens (1988:230) indicate that the Klahhane Ridge goat population consisted of 181 individuals in 1977 and 222 individuals in 1980. The density of goats should have gone up, according to their figures, rather than down, according to the figures of Olmsted and Pfitsch and Bliss. But note that Pfitsch and Bliss's population estimate of 180 goats is significantly different than Houston and Stevens's 222 goats. Pfitsch and Bliss's work was funded by Olympic National Park and not published until early 1985, so I cannot explain why they did not cite the park's estimate of the mountain-goat population (Olympic National Park 1984). While the discrepancies here underscore some of the problems with the flow and exchange of scientific data, there are more important matters to consider.

Pfitsch and Bliss (1985:109) write that "despite the high densities of mountain goats on Klahhane Ridge, indications of severe damage to their habitat by overgrazing, such as the prevalence of unpalatable species, are lacking. The most obvious effect of the goats on their habitat has been the creation of dust baths (wallows) which have become quite extensive." The critically important observation that evidence of severe damage to the habitat is lacking contradicts nearly every other statement written about the plant communities of the Olympics. Pfitsch and Bliss found some evidence indicating that grazing by mountain goats did in fact modify plant biomass and ground cover. For example, in "exclosures" (which generally consist of a fence built around a patch of vegetation in order to exclude herbivores from grazing or trampling on the plants) "total plant cover was the same" as on adjacent unexclosed plots, but the relative abundance of some species differed significantly between unexclosed and exclosed areas (Pfitsch and Bliss 1985:112).

Pfitsch and Bliss (1985:115) indicate that "most plant species on Klahhane Ridge showed evidence of grazing" during their 1978 through mid-1982 study. They also found that the "net annual production inside and outside of exclosures in 1980 was not significantly different," but that species apparently preferred by goats as forage tended to have lower production outside exclosures than inside and that other species "showed no consistent pattern" of productivity (Pfitsch and Bliss 1985:117). Of the 51 plant species they studied, "25% [13 species] were classified as preferred" foods for mountain goats (Pfitsch and Bliss 1985:115).

Pfitsch and Bliss (1985:118) note that "heavy grazing may be reducing the importance" or biomass of two heavily grazed species, *Festuca idahoensis* and *Carex spectabilis*. The former species made up significantly more of the ground cover within five-year-old exclosures than on adjacent areas. Such changes were "*presumably* due to the removal of biomass by grazing and to the detrimental effects of such removal on plant growth" (Pfitsch and Bliss 1985:120; emphasis added). They had evidence that the plants that had lower productivity were being grazed, but they were cautious in concluding that the cause was simply overgrazing. Such caution is rare in reports concerning the Olympic mountain goats.

Some of the material reported by Pfitsch et al. (1983) duplicates material in Pfitsch and Bliss (1985). But there is also some additional information. Pfitsch et al. (1983) found that plant-species density and plant cover were less on Klahhane Ridge than on one of two areas with low goat density (they did not report the results for the other area). They also found that "about 34% of the Klahhane meadow is bare," in contrast to 9% of a meadow with low goat use (Pfitsch et al. 1983:10). It is important to note that Pfitsch et al. (1983:13, 22, 98) did their studies of plant productivity and biomass during years when precipitation was less than normal and winter snowpack was only 6–25% of normal; many plants thus experienced "water stress." They therefore suppose that plant productivity was lower than normal when they did their study. The effect of this is not at all clear, but it might magnify the impacts of grazing by goats on plant productivity.

Comparisons of exclosed and nonexclosed areas on Klahhane Ridge indicated less plant cover in grazed areas than within exclosures. Further, Pfitsch et al. (1983) suggest that as disturbance by goats continued the density of resident, noncolonizing plants would decrease while the density of colonizing plant species would increase. Mosses that cover the ground and increase sediment stability (decrease erosion) would decrease in abundance as well. Pfitsch et al. (1983) indicate that revegetation to pregoat conditions may require active seeding and transplanting efforts by park personnel. "Only with long-term monitoring of the sites established for measurement of soil erosion, plant recovery, and plant production can [the] predictions [regarding impacts and recovery] be confirmed or refuted" (Pfitsch et al. 1983:103). Here again we find due caution. But there are no readily available data suggesting that long-term monitoring has taken place.

Some typed notes from a March 1987 presentation to park staff made by Edward G. Schreiner, a botanist who has spent many years studying the interaction of plants and goats in the park, indicate that there was no "clear cut evidence" that one of the plants critical to the controversy—Cotton's milkvetch (*Astragalus cottonii*)—was in danger of being extirpated by mountain goats (Warren 1987). Further, "no other endemic [plant] is endangered by goat impact." Finally, the notes indicate that "marmots and deer eat some of the same foods [as goats], but a lot of food is present. Some cases of competition have been observed where goats have driven deer away. Also where goats have been removed deer are returning. But no hard data [have] been collected. These are intuitive conclusions" (Warren 1987). These statements indicate a paucity of relevant data and that no plant form endemic to the Olympics was at that time in danger of being extirpated by the activities of mountain goats. Does an unpublished but occasionally referenced report by Schreiner (1987) contain similar statements? Where are the data clearly indicating the serious and significant impacts of the mountain goats on the alpine ecosystem?

RECENT STUDIES

The last chapter of Schreiner et al. (1993:126), a report on mountain goat–plant interactions commissioned by Olympic National Park that has on its cover the words "NOT FOR PUBLICATION," notes that "plants and herbivores are inescapably linked." It is clear that if a herbivore is present in an ecosystem it will influence the population size, biomass, and growth and reproduction of the plants it eats and tramples. The preceding chapters in Schreiner et al. (1993) document the interaction of plants and mountain goats in Olympic National Park.

The report begins by noting that the authors will test two general null hypotheses:

1. "Reducing mountain goat densities will not result in changes to the relative abundance of plant species (i.e., plant community structure)," and
2. "Mountain goats pose no threat to the long-term persistence of rare plants [in Olympic National Park]." (Schreiner et al. 1993:2–3)

It seems quite likely that the first hypothesis would be rejected simply because mountain goats are herbivores and thus modify the abundances of the eaten plant species as well as the noneaten plant species, which, due to reduced competition, could increase in abundance. If it is found that mountain goats somehow lower the biomass of rare plants, then the second hypothesis would also be rejected. Schreiner et al. never explicitly state their alternative hypotheses:

> **Alternative 1.** Reducing mountain goat densities *will* result in changes to the relative abundance of plant species (i.e., plant community structure).
>
> > **Corollary A.** Reducing mountain-goat densities will result in increased abundances of native and endemic plants and the reappearance of natural or native plant community structures.
>
> **Alternative 2.** Mountain goats *do* pose a threat to the long-term persistence of rare plants in Olympic National Park.
>
> > **Corollary A.** Reducing mountain-goat densities will result in the long-term persistence of rare plants in Olympic National Park.

There is, of course, no way to test—and thus no way to falsify—either of the alternative hypotheses or their respective corollaries until perhaps a hundred or more years have passed. Rather than review all of the data presented by Schreiner et al. (1993), I will briefly consider only their conclusions regarding the two null hypotheses.

The first null hypothesis is rejected (Schreiner et al. 1993:73) because in one of three areas studied the researchers found a negative statistical correlation between the density of mountain goats and the amount of plant cover: the greater the density of goats, the less the plant cover. Interestingly, that one area is Klahhane Ridge, where the population of mountain goats was denser than anywhere else in the park.

The researchers also reject their second null hypothesis (Schreiner et al. 1993:87, 88), even though they note that the grazing damage they found only "*appeared* to be from mountain goats" (emphasis added). But the sentences preceding and following rejection of the hypothesis are revealing, noting that "the long-term persistence of the *rarest* [plant] species *may be jeopardized* as the result of increased disturbance caused by mountain goats" and that "data presented here *cannot* provide direct evidence of whether or not goats will drive rare plants to extinction" (Schreiner et al. 1993:88; emphasis added). Is the second hypothesis rejected because any organism that feeds on another organism poses a potential threat to the "long-term persistence" of the latter as a species? If so, then phrasing the null hypothesis this way in the first place denies the constantly changing interrelations of organisms within an ecosystem (e.g., Krebs 1988).

To their credit, Schreiner et al. (1993) attempted to strengthen their rejection of the second null hypothesis by studying the impact of mountain goats on one particular rare "endemic" plant form, the Olympic Mountain milkvetch (*Astragalus australis* var. *olympicus*). They recorded all injuries to individual plants of this form that resulted from trampling and grazing over a period of seven years. They indicate that some instances of damage could not be clearly attributed to mountain goats (Schreiner et al. 1993:93), but nonetheless include all instances in their analysis. This, of course, will make the goats appear to be much worse ecological villains than they might really be.

Unfortunately, relevant data are not presented in a way that would allow evaluation of this inclusion (Schreiner et al. 1993:96). For example, we are told that certain sets of vegetation sample plots have "wallows" within fifty-five yards of them, but we do not know if each individual sample plot has a wallow within fifty-five yards. Similarly, we are told that twenty-two instances of goat feces were found in sample plots, but we are not told how many plots contained those feces.

In brief, Schreiner et al. (1993:99–100) indicate the following:

1. A longer-term study is needed to determine the population trends of the milkvetch.
2. Milkvetch and mountain-goat ranges "*appear* to substantially overlap" (emphasis added).
3. The typical density (number of individual plants per unit of area) of milkvetch is unknown.
4. The "overall long-term effects goats have had on [milkvetch] population dynamics" could not be determined.
5. Four individual milkvetch plants out of a population of 1,442 were killed by goat activities over a seven-year period.

Schreiner et al. (1993:100) reject the hypothesis that mountain goats pose no long-term threat to the survival of the Olympic Mountain milkvetch. This is absurd. How could the death of 4 plants in a population of 1,442 (or 0.3%) over a span of seven years pose a significant threat? Of course the mountain goats pose a *potential* threat, but so do humans, marmots, deer, elk, and the vagaries of climatic change. Schreiner et al.'s admission that they do not know what the long-term effects of mountain goats on the milkvetch might be invalidates their rejection of the hypothesis.

Ecosystems constantly change and develop, and the organisms of which they consist evolve as climatic patterns change. This is not meant to excuse the goats, but to underscore the implications of rejecting the null hypotheses. Null hypotheses are set up to be falsified and rejected. The lack of explicit statement of alternative hypotheses, however, and the fact that wording of these hypotheses may preclude their being tested suggest a hidden agenda: to find evidence that the mountain goats are having serious adverse impacts on the alpine ecosystem. This is clear in Schreiner et al.'s (1993:128) last statement: "Mountain goats are likely to have a far greater influence on individual plants, communities, and ecosystem processes than we have demonstrated here." This is difficult to imagine, given the seven years' worth of data they have mustered and the fact that the goat population has apparently remained stable, if not in fact decreased, since the completion of their field research in 1992.

Many, but not all, of the observations, data, and conclusions of

Schreiner et al. (1993) are included as six chapters and four appendices in a monograph authored by Houston, Schreiner, and Moorhead (1994). Schreiner is the author or coauthor of all the chapters.

As in the unpublished version (Schreiner et al. 1993), the first null hypothesis is rejected: "The null hypothesis that plant relative abundance would not respond to changes in mountain goat density was rejected" (Schreiner 1994:138, 172). Again, this is a predictable result, and I wonder how the alternative hypothesis reads.

The Olympic milkvetch was chosen as representative of the rare endemic plants that are consumed by mountain goats probably because it "is listed as threatened by the Washington State Natural Heritage Program as well as a Category 1 candidate for the federal endangered species list" (Schreiner 1994:179). As in the unpublished report, we are told that "goat wallowing in three plots was responsible for four dead plants in 1986" (Schreiner 1994:180). But "these preliminary investigations of rare plants in the presence of mountain goat herbivory were inconclusive"; the scientists "were unable to quantify the risk mountain goats pose to the persistence of rare-plant populations"; and "goat effects on [milkvetch] were sporadic, sometimes intense, and had unknown long-term consequences to the survival of the species" (Schreiner 1994:184). Quite unexpectedly in light of the unpublished report (Schreiner et al. 1993), we read in the published version that "we have insufficient data to cleanly reject the hypothesis that mountain goat herbivory does not jeopardize the persistence of rare plant populations" (Schreiner 1994:184).

Despite the lack of rejection of the second hypothesis, the final pages of the published version of the report (coauthored by Schreiner and Houston) state that mountain "goats are generalist herbivores with the capacity to consume most plant species, including rare plants"; thus the biologists "remain concerned about the long-term persistence of rare plants in mountain goat habitat" (Schreiner 1994:188). They underscore this concern by citing cases where rare endemic plants elsewhere have been driven to extinction or near extinction by introduced herbivores.

This concern for rare plants makes for wonderful media if one wishes to characterize the mountain goats as bad guys. For example, the Draft EIS (Olympic National Park 1995:8) states in one paragraph that the Olympic milkvetch is a rare endemic that is grazed by goats; it is a category-1 candidate for the federal endangered species list; it is listed as threatened by the Washington Natural Heritage Program; and, finally, it has been documented that wallowing by mountain goats killed individual milkvetch plants. These facts are often found strung together in just this fashion in the popular literature (Burger 1987; Flanagan 1995; Luoma 1993; Sleeper 1990; Sleeper and Hutchins 1989; Weingrod 1994; Wright 1996). Reading this statement clearly leads to the conclusion that the mountain goats are the reason the milkvetch is officially listed as endangered.

Milkvetch is not listed as threatened because it is eaten and trampled by goats, however. It is listed because it occurs in an extremely limited geographic area of the Olympic Peninsula.[3] Its range of occurrence is so small that it is more susceptible to extinction than are more widely distributed plants. Climatic change, infestation by fungus, predation by bugs, grazing and trampling by mountain goats, and trampling by people are all threats to the survival of this plant (Kaye 1989). Goats do occasionally kill milkvetch plants. But the threats of bugs, humans, and climate are never mentioned by park personnel.

Ironically, the significance of the Olympic milkvetch to the Olympic National Park mountain goat controversy was undone in February 1996. The *Federal Register* (61, no. 40:7609) removed this species from candidacy for threatened or endangered status. Apparently, this plant is not nearly so rare as Schreiner's (1994) research suggests and is in fact rather abundant and also widespread.

EXCLOSURES AND EMOTIONS

A commonplace technique for studying the impacts of grazing herbivores on vegetation involves the use of exclosures. Vegetation within the exclosure is compared over time with vegetation outside of the exclosure in terms of species represented, amount of plant growth, and other variables in an attempt to determine the impact of grazing animals on the vegetation. Exclosures have been used in the Olympics to study the influence of mountain goats on local vegetation ever since Olmsted's (1977, 1979) pioneering studies (Olympic National Park 1984, 1987a; Pfitsch et al. 1983).

[3] Most of the information in this paragraph was provided by Keith Gunnar, who talked extensively with John Gamon, botanist for the Washington Natural Heritage Program (WNHP). WNHP lists the Olympic milkvetch as threatened, but there are no regulatory requirements for such listing because there is no state law concerning threatened or endangered plants. Thomas Kaye (1989:65) has documented that a species of the weevil genus *Tychius* is the "chief seed predator" of the Olympic milkvetch, but he is unsure whether weevils, humans, or goats pose the greatest threat.

"Data derived from [exclosures] may be difficult to interpret" (Woodward et al. 1994:97; see also Macnab 1985:407–8). Reasons for this may include the small size of the exclosed patch relative to the ecological forces that drive vegetation dynamics over large areas; too few exclosures to provide a representative sample; inappropriate placement of exclosures and nonexclosed sample plots within the vegetation mosaic; modification of microclimate by fence construction (e.g., soil moisture and temperature regimes modified, air currents altered, amount of direct sunlight lessened by shade from fences or increased by greater grazing outside of fences, variation in competition between exclosed plants and nonexclosed plants); and attraction of herbivores to the fence structure, which may exacerbate trampling, trail formation, and grazing intensity (Woodward et al. 1994:97, 104). But park researchers also state that exclosures "can provide insight into ecological processes if the information is interpreted carefully" (Woodward et al. 1994:98; see also Schreiner et al. 1996).

Seventy-nine exclosures were constructed in the park between 1977 and 1981, inclusively (Olympic National Park 1984:50). Sixty-four (81%) of these enclosed 1 m^2 or less; other exclosures ranged in size from 16 m^2 to 100 m^2, with an average size of 48.8 m^2. Olympic National Park biologists (1984:60) indicate that the minimum size of exclosures recommended by other biologists is 1,000–4,000 m^2; thus the first problem, vegetation patches that are too small, is evident in the park data. Biologists (Olympic National Park 1984:51–60) studied only two exclosures (each 10 × 10 m), leading to the second problem listed in the preceding paragraph: nonrepresentative samples. The fourth problem, artificial modification of microclimate, also finds expression in the data: of the ten plant species monitored within the two examined exclosures, three seem to have responded to decreases in goat population size, two responded to climatic variation including microclimatic effects of the exclosure structure, two responded to both of these factors, and three displayed no response. Thus, park biologists (Olympic National Park 1984:60) indicate that "it can be very difficult to determine the causes of change within exclosures."

Difficulties in interpreting the exclosure data are exacerbated by the fact that the park biologists were comparing vegetation within exclosures to unexclosed sites "with a decreasing goat population rather than [unexclosed] sites with a more or less constant goat density" (Olympic National Park 1984:61). They suggest that "vegetation responded positively to exclusion of goats....Complete removal of goats (by fenced exclosures) had the most beneficial effect on vegetation" (Olympic National Park 1987a:25). However, they present none of the data necessary to substantiate this conclusion. Determining whether goats were actually responsible for the perceived changes is also hindered by the fact that there is dietary overlap among mountain goats, marmots, and deer in the area where the exclosures were constructed (Olympic National Park 1987a:44–45). Thus, it seems that mountain goats are (or were, prior to the reduction of the population in the 1980s) having an impact on the alpine vegetation of the Olympics, but the magnitude of that impact is not at all clear. More recent studies could not quantify the impact either (Schreiner 1994).

What is the public told regarding the exclosure data? The first issue of the *IGMT Newsletter* (1:5) states that "goat numbers must be reduced by at least 80% before some native plant communities can appreciably recover." The second issue of the *Newsletter* (2:5) states that "study plots inside fenced exclosures showed increased plant cover compared to areas open to goat grazing"; that "permanent plots monitored from 1979–1991 have shown changes in plant composition and cover, including some plant recovery after goat populations were reduced"; and that a 90% reduction in the goat population allowed some (park-desired) plant species to increase and weedy species to decrease. Barbara Sleeper and Michael Hutchins (1989:20) state that exclosure studies indicate that one species of plant lost 60% of its biomass to grazing by goats. Janis Burger (1987:29) reports that vegetation within exclosures recovered "from earlier damage" by goats. Nowhere in these widely published and circulated documents is the public told about the weaknesses of the available data, the difficulties of interpreting exclosure data, or the fact that other animals might have been responsible for some of the perceived differences.

In the process of simplification, significant observations and due scientific caution have been abandoned. Interpretations of the exclosure data are presented as solidly established scientific facts in widely read magazine articles and newsletters. These interpretations suggest an attempt to muster public support for the contention that it is not only correct but scientifically wise to significantly reduce, if not totally eradicate, the mountain-goat population in Olympic National Park. If this procedure is typical of scientific research informing management decisions of the late twentieth century, then we have a serious problem that exceeds the mere "muddling through" (Bailey 1982) of modern wildlife management and enters the arena of psychological or sociopolitical manipulation (see chapter 5).

HISTORICAL-PHOTOGRAPH PAIRS

Some of the most readily understood documentation of the *possible* effects of goats on the alpine plants of the Olympics is a series of historical-photograph pairs included in two of the park's unpublished reports (Olympic National Park 1984, 1987a). The 1982 management report indicates that

> thirty early photographs of the Klahhane Ridge area were collected [in 1981] and vegetative changes will be assessed by rephotographing earlier scenes as close to the same season and time of day as possible. Photos taken from 1889 to 1933 (N = 6) show the vegetation prior to the introduction of goats. Photos from 1933 to 1970 (N = 24) document the nature and rate of change that has occurred. This information should help provide a baseline for restoration of native plant communities.... To date [March 1982] ... about one-half of the scenes have been rephotographed. (Olympic National Park 1982:38–39)

Two things are worthy of note here. Of the six 1933 and earlier photographs mentioned above, only three have been included in later reports; nine of the 1933 to 1970 photographs have been included. Second, Harry Majors (1981a, 1981b, 1981c, 1981d, 1984a, 1984b, 1984c) reprinted many (if not all) of the photographs taken by members of the 1890–91 *Press* Expedition, which traveled up the Elwha River into the mountains and then came down the Quinault River. Although none of these are of Klahhane Ridge, this is certainly a data source that could be exploited by park researchers but apparently has not been.

Photograph Pairs in Park Reports

The historical-photograph pairs produced by park biologists typically show the same area twice, once in the 1940s–60s and once in the 1970s–80s. As the park biologists acknowledge, using these photograph pairs as evidence of the impacts of goats on the ecosystem is flawed from the start. "Since conditions prior to the introduction of goats cannot be well documented, the effects of these animals have to be determined in a reverse fashion; e.g., by studying the response of plants and soils to the removal of goats" (Olympic National Park 1987a:23, 25): because erosion and plant loss slow once goats are removed, then the goats must have initiated and contributed to the erosion and the plant loss. The flaw here is that correlation of goat removal with the slowing of erosion and plant recovery indicates that goats probably contributed to erosion and plant loss but does not mean that mountain goats *initiated* either. The impacts are not only not well documented, they are unknown and unknowable because "the original (pre-goat) vegetation of Klahhane Ridge is unknown" (Pfitsch et al. 1983:7).

The captions to the photographs suggest a hidden agenda. Two of the reports in which the photograph pairs are found are unpublished. It is significant that I was unable to find copies of the first four management reports (Olympic National Park 1981, 1982, 1983, 1984) in university libraries in the Pacific Northwest. The three most recent of these reports are prominently marked on their first page with the words "NOT FOR PUBLICATION" in capital letters. The 1981 report "was released for public review in February" of that year (Olympic National Park 1982:44), but this management plan contained minimal data and few analyses and conclusions that could be evaluated. It is not clear how it was distributed, who it was distributed to, how many people read it, or how many people provided comments about it to Olympic National Park managers.

I did find a copy of the 1987 management report (Olympic National Park 1987a) in the Oregon State University library in the fall of 1993 and a copy in the University of Missouri–Columbia library in February 1994. Comparison of the discussion and captions in the 1984 (unpublished, uncirculated) report with the discussion and captions in the 1987 report (which, while not published, was distributed to at least some university libraries) reveals some startling differences. The 1984 captions tend to be cautious, but the 1987 captions clearly implicate the mountain goats as ecological miscreants.

A synopsis of the historical-photograph pairs is provided in table 4.1. Several photograph pairs appear in both the 1984 and 1987 reports. In the 1984 report the biologists present four pairs of photographs and indicate that they "represent some of the changes that have taken place since ca. 1965 on Klahhane Ridge. Changes in [plant] species composition are apparent, but *cannot always be attributed to mountain goats*" (Olympic National Park 1984:74; emphasis added). They note that the erosion and changes in plant species shown by one photograph pair (1984 plate 2) are "obviously related to mountain goat disturbance but human trampling is undoubtedly responsible for some of the changes since this is a favorite lunch and resting spot of visitors. Thus, while people may visit the area to

TABLE 4.1.
Summary of Historical-Photograph Pairs Included in Two Unpublished Park Reports (all are views of portions of Klahhane Ridge)

Photo pair	Goats in photo?	Date of photo	Olympic National Park figure caption summary
		Olympic National Park 1984	
Plate 1a Plate 1b	present absent	~1965 Aug. 12, 1981	shows loss of plant cover and soil due to goat use
Plate 2a Plate 2b	present absent	July 20, 1970 July 15, 1984	shows loss of plant cover, plants, and soil "probably" due to goat use
Plate 3a Plate 3b	absent absent	Sept. 1967 Aug. 21, 1981	shows loss of plants and soil; goats "believed" to have started erosion and "probably" exacerbated continuing wind erosion
Plate 4a Plate 4b	absent absent	Aug. 27, 1970 Aug. 12, 1982	shows loss of western wormwood, "cause unknown"
		Olympic National Park 1987a	
Plate 1a Plate 1b	absent absent	Aug. 27, 1970 Aug. 12, 1981	shows "substantial" loss of western wormwood due to "trampling and grazing" by goats
Plate 2a Plate 2b	absent absent	Oct. 20, 1981 Oct. 7, 1983	shows invasion of weeds into wallow
Plate 3a Plate 3b	same photo as 2b absent	Oct. 7, 1983 Sept. 24, 1985	shows invasion of weeds into wallow and slow recovery after goat use stopped
Plate 4a Plate 4b	absent absent	~1963 Aug. 26, 1982	shows "considerable" loss of plants and soil "due to goat grazing and trampling"
Plate 5a Plate 5b	present absent	~1965 Aug. 12, 1981	shows loss of plant cover "due to" goat use and a wallow "created by goats"
Plate 6a Plate 6b	present absent	Aug. 1970 July 16, 1985	shows increased plant cover and recovery of a talus slope after goat population reduced
Plate 7a Plate 7b	present absent	July 20, 1970 Aug. 12, 1981	shows loss of plant cover and recovery of exposure of soil probably due to goats
Plate 8a Plate 8b	absent absent	Sept. 1967 Aug. 1974	shows loss of vegetated turf and expansion of a goat wallow
Plate 9a Plate 9b	same photo as 8b absent	Aug. 1974 Aug. 21, 1982	shows loss of vegetation and increased exposure of rocks due to erosion in an "extensively used" goat wallow

observe mountain goats, it cannot be stated that goats are directly responsible for all the increases in [erosion and a plant species that favors disturbed areas]" (Olympic National Park 1984:74). Thus, in 1984 the park biologists were cautious.

The 1984 plate 2 is reproduced as plate 7 in the 1987 report. The text does not refer to this plate, but the captions are identical, except for one added sentence in the 1987 report: "The disturbance-oriented species, common yarrow, is more evident."

But not all the captions are similar in the 1984 and 1987 reports. For example, compare the captions for the 1984 plate 1 and the 1987 plate 5, which are identical reproductions of the same photograph pair:

1984	1987
• Foreground shows a general loss of plant cover due to mountain goat use.	• Foreground shows a general loss of plant cover due to mountain goat use.
• Left of center shows an increase in exposed bare mineral soil, and perhaps some soil loss as the result of goat wallowing.	• Left of center shows a wallow created by goats and an increase in exposed bare mineral soil. Some soil loss may have occurred. Considerable numbers of plants have been eliminated.

Between 1984 and 1987 we have shifted from erosion that is "perhaps" the result of goat wallowing to erosion that was without question "created" by goats. Further, in 1987 the photographs purportedly show that "considerable numbers of plants have been eliminated" (this might result from more intensive scrutiny of the photograph pair and somehow tallying the number of individual plants shown in each photograph).

One of the most startling differences is in the captions for the 1984 plate 4 and the 1987 plate 1, which show identical photographs:

1984	1987
• Foreground and left foreground shows loss of most *Artemisia ludoviciana.* The cause of this is basically unknown. A climatic change could be responsible or perhaps in creased goat disturbance. This is a	• Foreground and left foreground shows loss of most western wormwood. The site is regularly inhabited by goats. Trampling and overgrazing have produced this substantial reduction of plant

good example of a visible change where we will be unable to determine the cause.

cover. Western wormwood is a preferred goat forage species in this area.

The scientific name of western wormwood is *Artemisia ludoviciana*, so the first sentence of these two captions is the same. But we have inexplicably gone from never being able to determine the cause of the reduction of plant cover in 1984 to knowing that the cause is trampling and overgrazing by mountain goats. This change is not the result of discovering that western wormwood is a preferred forage species; park biologists knew that as early as 1980, when they listed this plant in a table of "plant species which are preferred mountain goat forage on Klahhane Ridge" (Olympic National Park 1981:26). Perhaps "climatic change" has been eliminated as a potential causal agent based on research that is not indicated, but I suspect that the figure caption was changed for a different reason.

The 1984 report was "NOT FOR PUBLICATION" and was not generally distributed, but the 1987 report was circulated. Reading the two captions would lead to different conclusions about the ecological impact of the goats and how they should be managed. The figure captions are *interpretations* of photographs provided by biologists who seek to reduce the goat population significantly if not totally eliminate it.

Plate 3 in the 1984 report is reproduced in the 1987 report as part of plates 8 and 9. Plate 3a is reproduced as plate 8a and plate 3b as plate 9b (see table 4.1).[4] The 1987 photographic pairs include an additional view. The captions read as follows:

1984

• East end of Klahhane Ridge and near where the trail heads north toward Lake Angeles. The site is referred to by some people as "The Big Blowout."

1987

• Goat wallows at the east end of Klahhane Ridge near where the trail heads north into the Lake Angeles cirque. The site is referred to by some as "The Big Blowout."

4 The latter correlation seems to be correct but may not be. The 1984 plate 3b is given a date of "Aug. 21, 1981," and the 1987 plate 9b is given a date of "Aug. 21, 1982." Both photographs are attributed to D. B. Houston and seem to have been printed from the same negative. The top of the 1984 plate 3a is more cropped than the 1987 plate 9b, but the twigs, sticks, stones, shadows, and all other details have identical locations in the two photographs.

• Foreground shows a general loss of plant clumps and perhaps some increased exposure of rocks due to soil loss. The edge of this wallow has receded. We believe that goats started this erosion by wallowing. The continued loss of vegetation and soil, however, is probably a function of wind erosion as well as additional wallowing by goats. Revegetation and soil stabilization would be required to stop erosion here–even in the absence of goats.

• Foreground shows general loss of plant clumps and expansion of a goat wallow.
• Center shows loss of large areas of vegetated turf and the general receding of the wallow edge.
• The area is used extensively by goats for wallowing.
• This blowout originated as a goat wallow. The continued loss of vegetation and soil, however, is a function of wind erosion as well as wallowing by goats. Revegetation and soil stabilization would be required to stop erosion, even in the absence of goats.

The descriptions and the statements concerning the continued modification of the area are rather similar. But the statements about the *origins* of the modifications are significantly different. In 1984 park biologists "believe" goats initiated the erosion, but in 1987 there is no doubt.

A statement in Pfitsch et al. (1983:33) is even more disturbing: "Wallows are found at the 'blowout' near Rocky Peak [near the east end of Klahhane Ridge], a site of anthropogenic origin that is maintained by wind ablation." Here *human* activities created the "blowout," which is perpetuated by wind erosion. This directly contradicts the 1987 figure caption.

We simply cannot trust the conclusions implied by the captions of the historical photographs in the 1987 report. Goats walk and wallow and eat plants, and to think they had no effect on the alpine ecosystem of the park would be folly. But what I find troublesome is the way the photographs are interpreted by the park biologists: goat grazing, trampling, and wallowing not only exacerbated but initiated the erosion. By 1987 there is no doubt that goats were the *only* culprits. No evidence to justify these changes in interpretation is provided.

The photograph pairs are significant for two other reasons. First, the early photograph in a pair sometimes contains goats, while the later photograph never does (see table 4.1). The subliminal suggestion is that goats are to blame for *all* illustrated differences between the photographs. Second, the dates on some of the photograph pairs reveal that most of the photographs were taken in the mid-year or warm months, but there is no consideration of the aspect or elevation of the area shown. If the 1987 plate 6 photograph pair is on a south-facing slope, then it is no surprise that the vegetation appears more lush and "recovered" as the caption indicates; the 1985 photograph was taken in July when vegetation could still be growing, whereas the 1970 member of that pair was taken in August, when vegetation on south-facing slopes had stopped growing. There even appears to be snow on adjacent north-facing slopes in the July 1985 member of the pair; no snow appears in the August 1970 member of the pair. There is no indication of the annual growing season for the years represented by the photographs, so we cannot determine if plants are more or less abundant due to climatic differences.

In the 1970 photograph there also is a single goat who is wallowing and raising quite a dust cloud in the process. The dramatic effect of this photograph is striking, and its implications for goat management should be clear to anyone. These photograph pairs certainly provide a persuasive set of images *suggesting* that the Olympic mountain goats are ecological troublemakers.

Photograph Pairs in Other Places

A single pair of photographs was published by Sleeper and Hutchins (1989:21). This pair of photographs is not included in either the 1984 or 1987 report (but it is in Schreiner 1994:170–71]; see below). The caption reads: "An area of Klahhane Ridge is shown in 1980, before the goats were removed by park officials, and in 1986, five years after removal." The 1980 photograph shows over a dozen goats in one place; there are, of course, no goats in the 1986 photograph. The difference in the vegetation shown in the two photographs is remarkable. Virtually no plants appear to be over a few centimeters tall within several meters of the goats shown in the 1980 photograph. The 1986 photograph shows numerous shrubs ranging from half a meter to a meter in height. The implication is clear: Get rid of the goats and the plants will come back. However, Sleeper and Hutchins do not indicate the month(s?) of the year when the two photographs were taken or indicate the climatic regime of the two years. Further, they do not note that in 1986 about 41 goats could still be found on Klahhane Ridge (Houston and Stevens 1988:230).

I am not accusing the authors of the article of failing to tell the truth.

I am simply pointing out that at first glance the historical-photograph pairs *seem* to provide clear-cut evidence of the impact of goats. Several factors should be kept in mind, however: (1) we really do not know what the illustrated place was like in pregoat times; (2) the goats were not observed to initiate or actually create all of the illustrated impacts; (3) the captions are *interpretations* rather than simple descriptions; and (4) relevant data are not provided to justify the interpretations. When one examines those photographs with this realization, their potential significance decreases considerably.

The pair of historical photographs used as plate 3a in 1984 and as plates 8a and 9b in 1987 was included in the January 1992 *IGMT Newsletter* (1:2). The caption reads: "A large area used as a wallow by goats on Klahhane Ridge in 1967 [this is 1984 plate 3a], and in 1985, showing increased erosion and bare ground." This is similar to the caption from the 1984 report cited above. But the 1992 newsletter caption does not mention the contribution of wind to the erosional process, as noted in the 1984 caption and by Pfitsch et al. (1983:33). Nor does it indicate that in 1984 it was only "believed" that goats had started the erosion; by 1987 that belief had somehow been converted to firm knowledge (see the caption quoted above for 1987 plate 9). Anyone reading only the captions in the newsletter would probably receive the message the park is seeking to convey: the goats alone are to blame for serious ecological disturbances.

The *IGMT Newsletter* (2:5 [June 1992]) reported that a "group of scientists met in Port Angeles to review Olympic National Park's research on the effects of mountain goats on park vegetation." On April 2 the "chairman of that committee" mentioned several highlights of the meeting to "interest groups and news media representatives." One of those highlights reads as follows:

> • Some pairs of photographs comparing goat habitat over time show very significant changes in vegetation during the time in which goat populations built up. Several photograph pairs from Klahhane Ridge show increases in plant cover after goats were removed from the area.

That is, the park has in its possession photographic documentation of changes in habitat that correspond in time with an *increase* in the goat population. Surely these are not the photographs included in the 1984 and 1987 reports, which illustrate only what happens when the goat population decreases.

Thus, a new variable has been added: an increase in goat numbers correlates with habitat degradation; but we are not privy to the evidence (the photograph pairs). While I do not deny that an increase in goats occurred, the wording of the figure caption makes the problem appear large. This newsletter, which was circulated to over 700 people and interest groups, does not mention that in 1984 park biologists were unsure whether the goats could be held responsible for at least some of the photographically documented habitat changes. It also does not indicate how it was determined that goats initiated the changes in habitat or whether all thirty historical photographs mentioned in the 1982 report were examined by the "committee of scientists" or only the nine pairs included in the 1987 report. If the committee members also read the 1987 figure captions, then they did not have a chance to form their own opinions about what was illustrated.

A Final Park Use of Photograph Pairs

Schreiner (1994:139–72) published thirteen pairs of historical photographs that were also included in the unpublished version of his report (Schreiner et al. 1993:113–25). Plate 11 in these reports is the one printed by Sleeper and Hutchins (1989) and discussed earlier; the captions contain essentially the same information. Plates 7 and 8a in Schreiner et al. (1993) appear as plates 8 and 9a, respectively, in Schreiner (1994) and as plates 5 and 1, respectively, in Olympic National Park (1987a). The information in the captions of these various reproductions is similar. Captions for the other plates are similar in Schreiner et al. (1993) and Schreiner (1994), with one exception.

Plate 5 in Schreiner et al. (1993:117) was reproduced as plate 6 in Schreiner (1994:156). The 1993 caption reads: "Though the phenology [the seasonal state of growth and development of the plant] seems similar, some of the differences in abundance of flowering plants may be due to climatic conditions or phenology." The 1994 caption reads: "Though the phenology seems relatively similar, there are some differences in the abundance of flowers." The change in the caption might be taken as evidence that the authors are simply being more cautious. But altering the caption to note the change in abundance of flowers also implies that the goats were responsible.

There is a significant difference between the caption for plate 5 in Olympic National Park (1987a:34) and plate 8 in Schreiner (1994:160):

1987a
• Foreground shows a general loss of plant cover due to mountain goat use. Left of center shows a wallow created by goats and an increase in exposed bare mineral soil. Some soil loss may have occurred. Considerable numbers of plants have been eliminated.

1994
• Foreground and middle ground: The whole area has less plant cover and much more churned up bare ground by 1981. The wallows on the left are much deeper, and new wallows developed along the left edge and toward the center. There is an increase in *Achillea millefolium.*

The captions are very similar except for the last sentence of each. Schreiner (1994:136) notes that *Achillea millefolium* (western yarrow) is a "disturbance-oriented" species that dominates areas that are heavily grazed by goats. As grasses were grazed and decreased in abundance and as the density of goats increased, western yarrow increased in abundance. Perhaps with closer study of these photographs (plate 8 in 1994 and plate 5 in 1987) Schreiner was able to identify western yarrow plants and draw his conclusion. In any case, it is clear that the 1994 caption is a much stronger condemnation of the mountain goats than the 1987 version.

Schreiner et al. (1993:103) and Schreiner (1994:139–40) indicate that "interpretations of differences between photograph pairs are conservative." Schreiner and his colleagues considered several factors that might have caused the changes in vegetation apparent in sixty-six photograph pairs (only thirteen pairs are reproduced in their reports), including fire, climate, mountain goats, and humans. They suggest that "modifications of vegetation and soils by mountain goats were visible even though goat effects were superimposed on those driven by climate, fire, and humans" (Schreiner 1994:140), but they do not tell us the distinctive visual signals they used to determine this. For example, when they state that changes in habitats seen in some photograph pairs "were attributable to goats in 12 instances, to natural events in 11, and to a combination of natural events and goats or people in 5" (Schreiner 1994:141), we have no idea how they made those attributions.

In some cases, such as comparisons of photographs taken in 1980–81 with photographs taken in 1986–87 that show increased plant cover, it seems that their attribution of this change "to reduced goat density" (Schreiner 1994:141) represents an inference of a causal relation on the basis of a correlation. Of course, a correlation does not necessarily signify a causal relation. I suspect that the weight of such correlations, the taxonomic identification of the plants, knowledge of plant ecology, and knowledge of mountain goat grazing habits led to these inferences. I can perhaps live with the captions—as long as it is recognized that they are *interpretations* based on such a reasoning process. The modification of figure captions through time is acceptable only if it represents a change in thinking due to accumulation of knowledge (early interpretations that may have been based on the first thirty rather than the total sixty-six photograph pairs eventually used, less knowledge of goat grazing habits when the first comparisons were made, etc.).

CONCLUSION

The plant data are often used to justify claims that the mountain goats in Olympic National Park are having significant impacts on the alpine ecosystem found there. Clearly, the mountain goats are influencing that ecosystem. What is critical—given NPS wildlife-management policies—is the magnitude of those influences. The fact that this magnitude is unclear has not halted the generation and use of factoids. As we shall see in chapter 5, the evidence is even weaker for other kinds of influences the goats might be having on park ecosystems.

Park managers agreed in the early 1990s that the "mountain goats *potentially* threaten plant communities, including some endemic species. Long-term studies are needed to determine *if* the goats are impacting the growth and distribution of vegetation in alpine and subalpine areas" (Peterson et al. 1994:732; emphasis added). This, of course, is not what the public is being told in documents such as the monograph on Olympic goat biology (Houston, Schreiner, and Moorhead 1994) or the Draft EIS (Olympic National Park 1995). Those documents, prepared by park personnel, indicate that the threat from goats is serious and that the growth and distribution of vegetation have been significantly impacted.

5

WALLOWS, PUBLISHING, FACTOIDS, AND EMOTIONS

My review thus far has necessarily been limited to the reports and papers to which I have had access. Part of the frustration with exploring the scientific issues in the controversy is that much of the evidence that the park biologists use to argue that the goat population must be eradicated is unpublished. I obtained several critical documents by requesting copies from the park office or from other scholars. But *all* relevant data should be widely available for examination by interested parties. Ideally, all relevant material would be published and subject to independent peer review. The Draft EIS on the controversy (Olympic National Park 1995) does not include all the relevant data and analyses. The two primary authors of Houston, Schreiner, and Moorhead (1994) write: "Mountain goat management at Olympic National Park has been controversial. No matter how the issue is eventually resolved, the decision must be based, in part, on scientific information. We trust that this monograph provides the information necessary" (Houston and Schreiner 1994b:12). Unfortunately, the monograph does not provide the information necessary for objective evaluation of the issues involved.

Finding published *statements* in the popular literature and management documents about the relation of goats and wallows is easy. Finding published *data* is an entirely different matter. In the absence of data, emotions begin to take over because objective evaluation is impossible.

WALLOWS

The unpublished 1987 management report contains a detailed statement of the park biologists' perspective on wallows and erosion, which I quote in full here:

> Goats expose mineral soil by foraging, trampling, bedding, pawing and wallowing. Wallows are among the most visible

> and significant consequences of the goats' presence in the park's high country. For purposes of grooming and relief from insects, goats churn up selected areas to take dust baths. With the resultant loss of plant cover, soil is lost to wind and water erosion. This condition is aggravated and often enlarged as goats return to the same wallow time after time. A small wallow originally suitable for only one animal may eventually become large enough to accommodate a group of animals. Goat wallows from 3 to 30 feet in diameter and up to 3 feet deep are common on Klahhane Ridge. From one such large wallow, 40 tons of soil have been displaced. The average loss from seven other wallows was 1.3 tons. More than 70 wallows were found in one 10-acre area on Klahhane Ridge. Wallows are common features in goat habitat elsewhere throughout the park.
>
> Lower goat density on Klahhane Ridge has been associated with some recovery of wallows.... It has been observed that use of a wallow by only one goat effectively prevents plant recovery. (Olympic National Park 1987a:45–46)

Pfitsch et al. (1983:32) provide additional information on wallows and erosion not found in park documents: "it is difficult to relate goat numbers and habitat usage to wallow numbers. Wallows can occur anywhere in prime goat habitat and have few environmental correlates other than the presence of a fine-grained soil mantle which implies a relatively stable surface. They are found in areas that goats utilize frequently and are thus common at higher elevations, close to rock outcrops, and on relatively steep slopes." It has been reported that mountain goats have created anywhere from 700-plus (Luoma 1993:39) to 1,000 ("Alien to Olympic Park" 1992) wallows within the boundaries of the park and that up to forty tons of sediment may erode from a single wallow (Sleeper 1990:21). Media coverage of the apparent effect of the goats on the alpine and subalpine ecosystem makes the problem sound serious. But it is difficult to determine precisely how serious it might be.

Early Studies

Studies of wallows in Olympic National Park have taken place in areas where "erosive rates are naturally high" (Pfitsch et al. 1983:5). More significantly, Pfitsch et al. (1983:37) conclude that "although it was not possible to quantitatively separate the erosion due directly to mountain goats from [that due to precipitation, ice formation, and runoff from melting snow], goats were implicated at all times during the snow-free season" (about half of the year). Goats were implicated because they were observed to displace sediment during wallowing and they reduce plant cover, thereby exacerbating erosional processes. But Pfitsch et al. (1983: 45, 47) caution that "the activity of mountain goats is but another factor in an already complex [erosional] system. Due to system complexity, it is probably not possible even if logically feasible to separate the erosion due directly to mountain goats from that resulting due to natural processes." At least some of their data are somewhat ambiguous and "may not represent the true relationship" between the amount of sediment eroded and goat-related variables.

Finally, Pfitsch et al. (1983:34) indicate that "many wallows initially marked in 1976 and last measured in 1980 appear to have little use [by goats] and to be revegetating." And this is at a time when the mountain-goat population is perhaps doubling in size (fig. 3.1) and in a place (Klahhane Ridge) where the density of goats is higher than anywhere else in the entire park! This directly contradicts the Olympic National Park statement quoted at the beginning of this chapter. Perhaps the park biologists are referring to a worst-case scenario rather than to overall trends.

Although it cannot be denied that mountain goats exacerbate erosional processes, it is also clear that "some meadow disturbance may have derived from black-tailed deer (*Odocoileus hemionus columbiana*), marmots (*Marmota olympicus*) or researchers (*Homo sapiens*)" (Pfitsch et al. 1983:10). This is intriguing in light of the fact that Pfitsch et al. (1983:6) suggest that "a large amount of disturbance is unequivocally identifiable as goat induced." One must ask what proportion of the total is attributable to mountain goats.

It is clear that marmots contribute to disturbance processes such as changes in plant communities and erosion (e.g., Del Moral 1984; Taulman 1989), but these factors are never mentioned by park personnel, perhaps because marmots are considered part of the natural ecosystem. Their potential contribution to disturbance processes is never factored into equations or analyses seeking to document the magnitude of the disturbance caused by mountain goats. The contribution of deer has not, to my knowledge, been documented, nor has it been mentioned by park biologists, who also tend to be silent about the contribution of humans.

Virtually all of the historical photographs discussed in chapter 4 were taken on Klahhane Ridge. The goats there became "accustomed to digging for minerals in [this] area where mineral blocks were placed for many years" (Stevens 1983:158–59). Olympic goats seem to need salt only on a seasonal basis (Olympic National Park 1982:45), a pattern documented for this species elsewhere (e.g., Hebert and Cowan 1971), as a response to seasonal dietary deficiencies. Although the practice of putting out salt blocks was discontinued in 1984 (Olympic National Park 1995: 22), "minerals that leached into the soil still remain" (Stevens 1983:36). Human urine deposited in the sediments of Klahhane Ridge may have partially replaced the mineral taste if not the nutrient source originally found in salt blocks. But the issue is more complex.

On Klahhane Ridge, "in the early 1970s, large groups [of mountain goats] were commonplace and easily attracted to human urine or salt as bait. A salt site was established in 1972 along the crest of the ridge. Many [goats] were captured at or near this site for research purposes" (Olympic National Park 1984:84). Areas where salt blocks were placed were "visited by the goats frequently. The vegetation in these places has been destroyed for the most part because of the heavy and frequent concentration of the goats in these areas" (Olmsted 1979:1146). According to Olmsted (1979), the use of humanly created salt sites resulted in heavier modification of the alpine ecosystem by the goats than any nonhuman-influenced or natural mountain-goat behavior.

Another study of the effects of placing salt on the landscape found greater than anticipated impacts on habitats (Henderson 1984), which probably led to the suspension of using salt bait in 1984 (Olympic National Park 1995:22). At one location goats apparently removed twelve cubic feet of sediment and reduced plant cover due to grazing and trampling (Henderson 1984:3). Interestingly, this early unpublished study also included data indicating that plant death might have occurred because of increased salt in the sediment rather than because of goat activities (Henderson 1984:4). This possibility is not mentioned in any of the research published by park biologists.

In the 1970s goats tended to be denser on Klahhane Ridge than in other areas of the park (Houston and Stevens 1988; Olympic National Park 1984; Pfitsch et al. 1983). Reasons for this may include the establishment of salt sites there, the large number of human visitors that use the trails along the ridge and urinate there, and the "learned behavior by goats to utilize a seasonal source of salt," though salt interest by goats "was initially evident on Klahhane Ridge in the mid-1960s, well before a salt site was established" (Olympic National Park 1984:91). The high density of goats in this area has also been attributed to the "rocky terrain for escape, open meadows for forage, and north-facing slopes for thermoregulation in summer" (Pfitsch et al. 1983:1). The park asserts that salting activities did *not* result in an increase in the density of goats: "studies of 130 tagged goats at Klahhane showed that goats using the salt were resident goats and salt did not attract goats from long distances" (Olympic National Park 1995:22). The source of data for the park's assertion is not stated.

Stevens (1983) showed that goats (she tagged 237 of them over five years) tended to disperse away from Klahhane Ridge in the late 1970s. She suggests that this was due to competition for scarce resources—a result of the dense goat population. But Stevens also notes that the goats were not studied intensively until the beginning of her research in 1977. In conjunction with Pfitsch et al.'s (1983:40) observation that the wallows appear to be only ten to fifteen years old, this suggests that significant wallow formation may have begun largely in the 1960s.

It seems reasonable to question whether the 1960s pattern of goat use of Klahhane Ridge was exacerbated by the intensive human use of the trails there, which can be reached by following a paved road from Port Angeles. The road and the trails afford excellent views of the alpine habitats in the Olympics. Even park personnel indicate that human activities exacerbated the impacts of goats on Klahhane Ridge, but their suggestion is cleverly worded: "It is possible that the public placed salt there to enhance photography or viewing opportunities" (Olympic National Park 1995:22). This is purely conjectural, but it implies that park personnel are free from ultimate responsibility for the damage.

The Klahhane Ridge example demonstrated that "after 10 years of goat use, a zone of intensively trampled vegetation and soil could be expected to occur at most new [salt] sites" (Olympic National Park 1984:91). Thus, after 1983 park biologists placed salt in rubber buckets at new sites and removed them in the fall; prior to 1983 they simply placed the salt blocks out on the landscape. In 1981 salt was left on Klahhane Ridge as bait for seven or eight weeks (Olympic National Park 1982:45). In 1982 "salt was placed as bait on Klahhane Ridge on June 8" and remained there for at least six weeks (Olympic National Park 1983:30). In these two cases, as in earlier instances, the salt was used as bait to attract mountain goats so that they could be captured and removed from the park. It is thus possible that the wallows are the end result of goats' pawing sediment to obtain salt minerals that were placed (either as mineral blocks or as urine) on Klahhane Ridge by humans.

Do mountain goats create wallows from otherwise undisturbed portions of the landscape, or do they find a place where sediment is exposed by some erosional process and exacerbate the process by using that place as a wallow? No one, so far as I know, has discussed this question, though park biologists have noted that goats "like areas that are already disturbed" (Warren 1987). But in the published literature we read about the number and size of wallows "created" and enlarged by goats in the Olympic Mountains.

Although wallows are not discussed or photographed as often as plants, they are often mentioned as a reason for the removal of mountain goats from Olympic National Park. Sleeper and Hutchins (1989:20), for example, make the following statement in a popular magazine:

> Dust-bathing has promoted considerable soil erosion. When grooming or when overheated or harassed by insects, the goats lie in the dirt and, with a quick movement of the foreleg, throw cool soil over their bodies. This behavior creates wallows—large patches of exposed soil that are habitually used by the goats. Studies of soil movement ... showed that up to forty tons of soil had been displaced from a single wallow. The area of the park with the highest density of goats contained over a hundred wallows in a single meadow.

Mountain goat wallows are surrounded by disturbance-oriented plants that have displaced "normal" community dominants.

Thus, not only is erosion exacerbated, but the composition of the floral communities is modified by wallowing. Note that the 100-wallows-per-meadow figure comes from Klahhane Ridge, where the goats occur in the greatest density. Sleeper and Hutchins do not mention why this particular area has such a high density of goats. Although one could conclude from close reading of the above quotation that fewer wallows are to be expected in other areas of the park where goat densities are lower, this point is not made explicitly. What sticks in one's mind is the 40 tons of soil per wallow (probably the same wallow mentioned in the quotation at the beginning of this section) and the 100 wallows per meadow.

Where Are the Data?

"Park staffers have found more than 700 wallows deeper in the park's interior" (Luoma 1993:39). This total does not include the wallows of Klahhane Ridge, where "more than 70 wallows were found in one 10-acre area" (Olympic National Park 1987a:45). Sleeper and Hutchins (1989) indicate that over 100 wallows were found in a single meadow on Klahhane Ridge. How many wallows are there in the whole park? How many are regularly used? Which animal species create and use them? Answering these questions is extremely difficult because data on wallows are seldom presented.

Burger (1987:30) indicates that in 1981 "park staff found more than 45 tons of soil displaced in [one] wallow." Is this the same one mentioned earlier as having lost forty tons of sediment or a still larger one? What proportion of the sediment is being displaced by the wallowing of mountain goats and what proportion by other processes? Burger (1987:30) states that "goats had displaced several hundred pounds of soil from the smallest measured wallow" on Klahhane Ridge; she does not mention that wind and water contribute to the erosion. Burger's statement is found in a widely distributed popular journal; the park biologists' statement that "soil is lost to wind and water erosion" (Olympic National Park 1987a:45) is found in the unpublished management plan.

Park biologists (Olympic National Park 1982:25) indicate that "studies of soil erosion in the meadows were initiated during the fall of 1981....Information on erosion in selected wallows will be useful, but a map of their distribution is also needed." In the uncirculated 1982 report (Oympic National Park 1983:16–17) they note:

> Because of the time required for plots to stabilize, intensive data collection will begin in 1983. Work on erosion and wallows must be continued since little information on soils disturbance has been collected to date....Wallows in the Big South Meadow area were numbered and measured by Olmsted in 1976. Wallows were remeasured in 1978, 1979, 1980 and 1982. Variability between investigators more or less precludes comparisons on an annual basis. Future work should be limited to wallows that are small and uniform in their dimensions—e.g., without irregular boundaries....Beginning in 1981, several wallows were measured with a transit. A pin placed at the top of each wallow serves as a permanent reference point. Results from measurements in 1982 will appear in [a] final report.

The final report referred to is the unpublished report of Pfitsch et al. (1983). Some of the results of this report are summarized in Olympic National Park (1984). It is relevant to note that Burger (1987) indicates that park managers knew in 1981 that *significant* erosion was taking place in wallows. How is this possible if in 1982 very few data on erosion were available? Where are the data on which the 40–45 ton figure is based? It is cited in two different places (Burger 1987; Sleeper and Hutchins 1989), so relevant data must exist, but I have been unable to find them.

What Do the Data Indicate?

The procedure used to measure rates of erosion on Klahhane Ridge in 1982–83, subsequent to Pfitsch et al.'s (1983) study, included placing sediment traps at the base of selected wallows and measuring the horizontal and vertical extent of individual wallows through time. Measurements were taken during the summer of 1982 and again in the summer of 1983. Because of difficulties with the procedure, data collected in 1981 could not be used in the park scientists' analysis. (It is unclear if this means that the 40–45 ton figure apparently derived in 1981 is valid or invalid.) The scientists found that erosion (the amount of sediment collected in the traps) was significantly less in 1983 than in 1982.

Park biologists write: "Reduced goat density could well be responsible for the majority of the difference in observed erosion rates.... While reduction in the overall goat population does not necessarily mean that a proportionate reduction in wallow use occurred, a substantial overall reduction in wallowing seems very likely.... Our *preliminary* interpretation is that most of the reduced erosion is associated with lower goat density" (Olympic National Park 1984:68; emphasis added). The mountain-goat population numbered 126 throughout the 52-day period when erosion measurements were taken in 1982. The goat population numbered 147 during the first 13 of the 108 days when measurements were taken in 1983, but only 85 individuals remained during the last 74 days. During the other 21 days, 52 goats were being trapped and removed (about 2.5 goats per day). To derive an average daily population for 1983, I multiplied 147 goats times 18 days, added that to 85 goats times 74 days, then added 116 goats (the approximate average population per day during the capture period) times 21 days, and divided the total by 108 days. The average daily population of goats during 1982 was 126, whereas the average daily population of goats during 1983 was about 105. Thus, perhaps the reduced density of goats is in part responsible for the reduction in erosion.

But there are significant problems here. Park researchers found that the amount of erosion was *not* correlated with the amount and intensity of precipitation during the period; they noted that these two variables are "usually correlated positively" (Olympic National Park 1984:65). Perhaps the amount of erosion that was measured was not in fact the amount that occurred. Or, if larger populations of goats cause more erosion, then perhaps the 1982 goat population was sufficiently large to override the effects of minimal precipitation-generated erosion, while the 1983 goat population was sufficiently small that even with more precipitation there was insufficient rain-generated erosion to surpass the amount of erosion in 1982.

While park researchers may have used some such reasoning in concluding that the goats were responsible for the erosion, they indicate that some of the difference in erosion amounts between the two years is probably attributable to their measurement techniques (Olympic National Park 1984:65, 68). Thus they conclude that "additional sampling will be required to clearly separate effects of goat density from climatic factors on erosion" (Olympic National Park 1984:68). By the park biologists' own admission, the data available in 1984 do not unambiguously indicate that goats are responsible for the magnitude of erosion that is regularly attributed to them. No records were apparently kept of the number of goats that used the wallows or how many hours the wallows were used during the measurement period. In this respect the research design is seriously flawed.

In addition, while wind is said to be an erosional process (Olympic National Park 1984:77), it is not mentioned in the discussion of erosion (Olympic National Park 1984:64–73). Yet we have Burger's (1987) statement and the implications of others (e.g., *IGMT Newsletter* 1:2; Luoma 1993) that goats are totally responsible for the measured erosion. Even the park researchers admit that these hypotheses need additional testing and refinement. Yet the statements concerning the studies of erosion express no doubts and seem geared toward reassuring the public that the park's desire to eradicate the goats is justified.

This is clear from the unpublished notes taken during a March 1987 presentation by park biologist Ed Schreiner to park staff: "In some places erosion is continuing so fast there is no recovery *even in the absence of goats.* This may apply to a portion of a wallow or to complete wallows" (Warren 1987; emphasis added). "In areas with a shallow mantle [thin sediment layer], wallowing readily displaces all of the fine sediment leaving a surface of scree which *ceases to be used by the goats*" (Pfitsch et al.

1983:25; emphasis added). This is not at all the impression one gets from the published statements of park researchers.

PUBLISHING

Stevens (1982) notes that the Olympic mountain-goat population did not become the subject of intensive study until 1977, when she began her research. As Joanne Michalovic (1984:139) notes: "National Park Service resource managers rarely have the luxury of implementing a pure research proposal when there are no resource problems. Investigations which do get extensive funding are usually justified upon the extent to which results can directly aid management in decision making, or in helping to solve a crisis situation....we cannot have the data collection and monitoring until the crisis arises." If this is the case in Olympic National Park, then the studies by Stevens (1982, 1983) and others were begun because of a perception in the mid-1970s that a crisis had arisen. The 1991 Resources Management Plan by Olympic National Park indicates that a Resources Management Plan drafted in 1970 identified the "problem of mountain goats ... as a threat to the integrity of park ecosystems." Study of that threat, however, initially concerned not just the effects of mountain goats on the alpine ecosystem, but the effects of human park users as well (e.g., Schreiner 1974, 1982). With few exceptions (Bell and Bliss 1973; Moorhead and Schreiner 1979), results of the research on human impact have not been published in popular or scientific journals and have not been publicized (see chapter 4).

The ecology of the mountain goats in Olympic National Park has been the subject of much recent research effort; but prior to the mid-1970s there was minimal concern for the issue and research was focused elsewhere. In figure 5.1 I have plotted the number of unpublished reports and published articles that directly concern the ecology or management of Olympic mountain goats per two-year period between 1950 and 1993, inclusively. This graph is based on a list of titles in the bibliographies of various documents, such as Olympic National Park (1981, 1982, 1983, 1984, 1987a, 1995) and Moorhead and Stevens (1982) (see the appendix). Because I have not seen some of the unpublished materials that contribute to this graph, it is possible that some titles have been inadvertently omitted. The few published statements that tend to raise questions regarding park management policies (Anunsen and Anunsen 1993; Lyman 1988a, 1989) are not plotted. The plotted articles and reports were

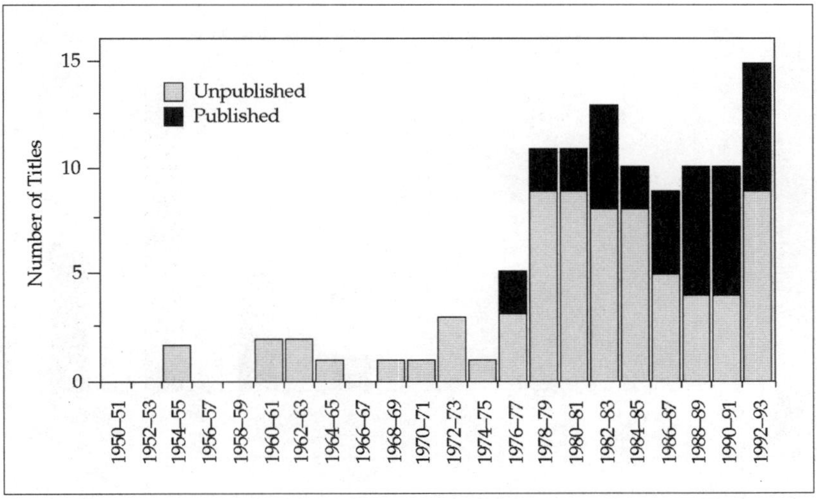

Figure 5.1. Frequency of reports and papers (published and unpublished) on mountain-goat ecology in the Olympic Mountains per two-year period between 1950 and 1993, inclusively (titles are listed in the appendix). Note the florescence of report production in 1976–77.

authored by park biologists or researchers and by others who interviewed park researchers. Note that the thirteen titles listed as appearing prior to 1976 (with one exception) concern estimates of the size of the goat population.

Beginning with titles released in 1976–77, we find reports and articles that deal with the ecology of the Olympic goats. I have distinguished between published (total titles = 35) and unpublished (total titles = 72) documents to illustrate that much of the evidence on which management decisions are said to be made has not undergone peer review, which tends to increase the quality of published papers and reduce errors (Maddox 1989). The most recent management policies of the national park system (USDI-NPS 1988) do not require park scientists to publish their research results in peer-reviewed journals, although perhaps they should. Some, but not all, of the originally unpublished information generated by Olympic National Park's researchers was later published; for example, two papers were extracted from one unpublished report (Olympic National Park 1984) and published in scientific journals (Houston et al. 1986; Houston and Stevens 1988). It is nonetheless apparent that the Olympic mountain goat problem was not perceived as warranting the serious attention of park researchers until the mid-1970s.

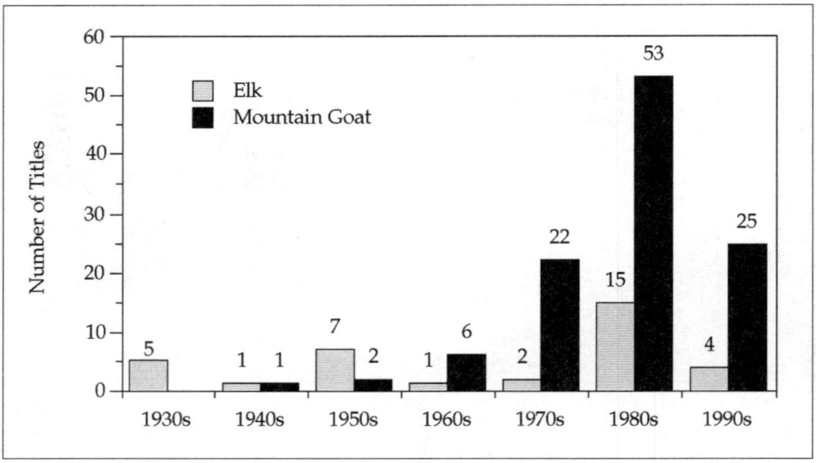

Figure 5.2. Frequencies of reports and papers on mountain goats and on elk on the Olympic Peninsula per decade (1990s = 1990 through 1993 only). While Olympic National Monument (the precursor of Olympic National Park) was founded in 1909 to protect the elk, note that the chief wildlife management concern of park officials over the past twenty-five to thirty years, in terms of report production, has been the mountain goats.

The numbers of reports and articles per decade beginning in the 1930s prepared on goats and Roosevelt elk are revealing. The Mount Olympus National Monument (the precursor of the national park) was created to protect the elk (Roloff 1934:226). Elk were studied more intensively than goats were early in the history of the park, but in the 1970s that trend was reversed: goats began to be studied more intensively than elk (see fig. 5.2). About 36% of the reports on elk were completed prior to 1960, while less than 3% of the reports on goats (3 of 109) were completed prior to 1960. Perhaps some poorly publicized detection of a crisis prompted the flurry of research on the goats in the mid-1970s (see fig. 5.1).

The frequencies of reports completed per decade for elk and for goats are correlated (Pearson's $r = 0.72$, $P = .067$). This suggests that the rate of research on both species has not been completely the result of a perceived crisis. Further, park managers began to remove goats experimentally in 1981 (Olympic National Park 1982), when only a few preliminary data indicating a crisis were available (Olmsted 1977, 1979). In the 1980s attempts to document the ecological crisis created by the mountain goats increased, apparently to justify the population-reduction activities already taking place.

FACTOIDS

By "factoid" I mean a bit of information that is repeated in multiple publications over the years and becomes accepted as a fact. The repeated information may, however, be merely a suspicion, a weakly founded hypothesis, or a bit of potentially significant data that in a particular context implies that the goats are a problem or should be removed. Whether or not use of such factoids is an intentional strategy that is meant to foster support for the park personnel's plan to eradicate the goats, the effect of factoids in the controversy is clear.

A Recent Synopsis of the Data

In a journal published by the NPS, Joseph Flanagan (1995:24) repeats Susan Schultz's (1993, 1994) observation that reports of goats in the Olympics prior to 1925 were made by "ex-soldiers, Indian fighters, prospectors, and cowboys" and that these observers "lack the specificity of a naturalist" (see chapter 6). He reiterates Schultz's critique of C. A. Barnes (1890) and S. C. Gilman (1896). Like Schultz, Flanagan fails to mention the natural-history errors made by late-nineteenth-century naturalists held in high esteem by park personnel. He repeats Schultz's (1993) observation that three (Schultz 1994 says four) individuals were "emphatic" about the absence of mountain goats and that D. G. Elliot's (1899) catalogue of wildlife "does not include mountain goats, even though they searched for the animal's tracks, droppings, and swatches of hair" (Flanagan 1995:24). This search for tracks, droppings, and hair is not mentioned by Schultz (1993, 1994) or by Elliot, so a new factoid is created.

Flanagan (1995:26) indicates that "investigators examined the archaeological record where mountain goats are indigenous, turning up *an abundance of remains*, even at coastal sites" (emphasis added). If this refers to an abundance of mountain-goat remains, it is simply not true. The zooarchaeological record for fourteen sites on the mainland coast of British Columbia and southeastern Alaska consists of 7,387 mammal bones and teeth that have been identified to at least genus; of these, only 83 (1.1%; from five sites) have been identified as mountain goat (see table 6.1 for a partial list). The zooarchaeological record from six sites in interior British Columbia in the heart of goat country produced 4,077 identifiable bones and teeth; 126 (3.1%) of these represent mountain goats.

Mountain-goat remains are not "abundant" in these archaeological contexts. Maybe Flanagan had some other context in mind, but his statement results in the creation of a new factoid.

Flanagan (1995:26) quotes Olympic National Park archaeologist Paul Gleeson as saying that we should expect mountain-goat bones in the Ozette faunal assemblage. Ozette, an archaeological site in the northwestern corner of the Olympic Peninsula, produced a large zooarchaeological collection (Huelsbeck 1983, 1994). But Gleeson's statement fails to appreciate Randall Schalk's (1993) conclusion that this site is "far from possible goat habitats," a point made by Flanagan (1995:24), and the fact that 94% of the identified mammal remains from Ozette represent marine mammals, a point not noted by Gleeson or Flanagan. The net result is yet another factoid.

Flanagan (1995) quotes two of the nine individuals who were paid to review Schultz's (1993) and Schalk's (1993) reports, who found the zooarchaeological and historical evidence convincing. He fails to point out that these individuals had not read the primary literature and thus lacked the knowledge to evaluate Schultz's review. Finally, Flanagan (1995:23) indicates that the Olympic Mountain milkvetch is grazed by goats and that it is "a prime candidate for the federal endangered species list." The implication that the milkvetch is a candidate because of goat grazing (and wallowing) is incorrect (see chapter 4). But another factoid created by park personnel is perpetuated by Flanagan.

I asked Flanagan, who is the associate editor of the journal in which his paper was published, if the journal would publish my rebuttal to his article. He informed me that this was not the journal's policy, but I could write a letter to the editors. Unfortunately, this is typical of popular journals that are meant to provide synopses of information to a wide readership. Thus, papers such as Burger (1987), Sleeper (1990), Luoma (1993), Flanagan (1995), and Wright (1996) have no published rebuttals. Even when rebuttals exist, as in the cases of Weingrod (1994) and Sleeper and Hutchins (1989), they are typically very limited in scope (e.g., Bishop and Martin 1994; Lyman 1989, respectively). The destruction of factoids is difficult.

Why Were Mountain Goats Released in the Olympic Mountains?

The mountain goats transplanted to the Olympic Mountains were obtained in exchange for elk (Webster 1932). Why goats were released is a potential point of contention. It has been suggested in public hearings on the controversy that the goats were originally obtained by hunters who wanted to shoot them, so animal-advocacy groups should not be overly concerned if the goats of the 1990s are eliminated by shooting (R. Anunsen, personal communication). But were the goats originally obtained for this purpose?

As noted in chapter 1, Webster (1920:146) indicated that "it is the intention of the Klahhane Club (Olympic Mountaineers), aided by the Forest Service, to introduce [mountain goats] as soon as it may be possible to secure them." A news story published in 1925 in the *Port Angeles Evening News* (and published that same year in the *Murrelet*) indicates that the release of the goats was orchestrated by Forest Service officials, Washington State Fish and Game officials, and members of the Klahhane Club ("Introduction of Mountain Goats in the Olympic Mountains" 1925). Twenty years later, Scheffer (1946:129) indicated that Washington State Fish and Game officials and Forest Service officials handled the release, a fact borne out by various historical documents (e.g., Morgenroth 1991:166). But what was the purpose of the release?

A newspaper article ("Mountain Goats May Be Introduced" 1919) provides the only more or less contemporary statement about the purpose of the release. This article reports that U.S. Forest Ranger Chris Morgenroth indicated in a talk given to members of the Klahhane Club on October 15, 1919, that the possibility of introducing moose, mountain sheep, and mountain goats to the Olympics "had first been brought up by the [U.S.] government biological survey." Schultz (1994:280) suggests that "for many years before their introduction, [Morgenroth] had advocated bringing either mountain sheep or mountain goats into the Olympics," but she cites no references. The newspaper account indicates that it was the U.S. Biological Survey that originally wanted these animals. During the October 1919 meeting Morgenroth appealed to the Klahhane Club and other mountaineering organizations in the state to "take up the necessary campaign of publicity to the end that a dozen or more of these animals might be brought here." The newspaper article concludes with the following sentence: "Eventually the animals would increase sufficiently, spreading throughout the mountains, so that an open season might be established" ("Mountain Goats May Be Introduced" 1919). Thus, the original goal *may have been* to produce a huntable population (see below), and it seems that the idea originated with the U.S. Biological Survey and perhaps with the U.S. Forest Service in the form of Morgenroth. Recent reporting on the origin of the idea and other facets of the introduction, however, does not fit these facts.

Two statements in the early literature suggest that sport hunting may *not* have been the only reason for the introduction of mountain goats. First, Webster (1920:153), president of the Klahhane Club, wrote: "The Mt. Angeles game sanctuary, established at the request of the Klahhane Club, was planned, in part, to provide a home for the Mountain Goat, *where it might live in no fear of molestation by human enemies*" (emphasis added). Clearly, at least initially the Klahhane Club was not soliciting support for the introduction of another game animal. Second, Scheffer (1946:129) indicated that "on January 6, 1925 [five days after the first goats were released], the Clallam County Game Commission created the Storm King Game Refuge *to protect the goats*" (emphasis added). Perhaps the plan was to protect the goats until their numbers were sufficient to withstand some hunting pressure, but that is not what is implied here. Only the 1919 newspaper article indicates that the goats were to become the target of hunters and that this was the intent of the U.S. Biological Survey. It appears that other interests at the time of the release wanted to protect the goats.

Park biologists' interpretation of these earliest statements about the release events is an exceptionally clear case of the use of factoids. Stevens and Driver (1978:165) indicate that the goats were released in "hopes of establishing another big game species on the peninsula for hunting purposes." They cite no references. An article three years later coauthored by Stevens states: "Why the mountain goats were brought to the Olympic Mountains is not a matter of record" (Hutchins and Stevens 1981:59). Had Stevens read the 1919 newspaper article, she would have known that it *was* a matter of record. The statement continues: "but since they have long been prized as trophy animals, the potential for establishing a harvestable population probably provided some incentive." Thus Hutchins and Stevens *speculate* that the goats were released so that they might eventually provide a huntable population. Within a few years this speculation would become a factoid.

In her 1983 doctoral dissertation, Stevens reports:

> The goats were ... introduced during a time when much of the [Olympic] peninsula was administered by the Forest Service as a National Monument and, therefore, hunting was allowed. When the Park Service took over the administration of the monument in 1933, they inherited a non-native population of

> mountain goats and brought with them the regulation against hunting. Although non-native animals are generally not welcome on Park Service lands, the population of goats was so small and at this time such a novelty that little attention was paid to goats and goat management, except to laud their beauty and advertise their occurrence as an attraction to Olympic National Park. (Stevens 1983:13–14)

This statement is incorrect on several counts. First, as noted by Webster and Scheffer in the citations above, the goats were not released in the area so that they could be hunted, at least initially, and hunting for goats was explicitly *not* permitted five days after the initial release.

Second, in 1905 the state of Washington made it illegal to hunt elk, in part to protect herds in the Olympics (Hult 1954:217–18), though miners could still kill them at any time (McLeod 1984:91). Limited elk-hunting seasons were permitted on the peninsula in 1933, 1936, 1937, and 1938 (Hult 1954; see also Houston et al. 1990; Twight 1983). Efforts to create a game refuge in the Olympics in 1906 and again in 1908 failed (McLeod 1984:104); but when President Theodore Roosevelt signed the law creating Olympic National Monument on March 2, 1909, it immediately became closed to both elk hunting and mining (McLeod 1984:106–7). Elk, the largest game animals on the peninsula, were overhunted to the point that the monument was created specifically to protect them. Some local groups with commercial and mining interests decried the fact that eastern politicians had created a refuge for "bear and other 'varmints'" and a "moose pasture" (quoted in McLeod 1984:109).

Many elk died over the winters of the second decade of the twentieth century, and as a result studies of the elk population were undertaken by the U.S. Biological Survey (Twight 1983:40). Robert Wood (1968:82) indicates that in October 1927 (not quite three years after the first goats were released) the Olympic National Monument "was closed to hunting and trapping in perpetuity in an effort to conserve the wildlife of the mountains." Ben Twight (1983:44–46), however, indicates that attempts to create a game refuge from 1926 through 1928 were unsuccessful, and later complaints by biologists about the dwindling elk herd went unheeded. Grazing by domestic animals was blamed for part of the loss of elk (Twight 1983:60). This has implications for the study of early (non-goat-impaired) vegetation communities and the creation of wallows (see chapter 4). But Stevens is only partially correct in suggesting that the

NPS administration of the monument/park initiated the no-hunting regulation in 1933. Hunting of elk had been regulated for twenty-five years prior to that date. And regulations against hunting goats had been in place virtually from the day of their release in 1925.

But what *was* the purpose of the release? Roger Hoffman and R. Gerald Wright (1987:18) state that "according to E. B. Webster [Anonymous 1925], a local journalist, H. M. Fisher, the county game commissioner and a group of local sportsmen were responsible for bringing the mountain goat to the Olympic area." Webster ("Introduction of Mountain Goats" 1925) does not mention Fisher by name (although Webster 1932 does) and does not mention any group of local sportsmen. Once again, a belief has become a "fact."

This "fact" is repeated often in the professional literature. Seth Tuler et al. (1992:318) write: "Sport hunters first introduced eleven or twelve mountain goats into the Olympic Peninsula in the 1920s." Tuler and Chuck Janda (1991:14) state: "Mountain goats were first introduced to the Olympic Peninsula by sport hunters in the 1920s." Flanagan (1995:23) writes that the drivers of the truck that transported the first goats in 1925 were there because of "hunting, and their riders—mountain goats—were meant to seed a population of game animals." Hoffman and Wright (1990:1) restate this in the published version of their report: "About 12 mountain goats were introduced to the Olympic Mountains by local sportsmen between 1925 and 1929."

Olympic National Park biologists initially simply reported the release events and did not suggest their purpose (Moorhead and Stevens 1982; Olympic National Park 1981, 1982, 1983, 1984). But Janis Burger (1987:28), a park employee, states that the goats "were introduced into the Olympic Range as game animals." That same year Olympic National Park (1987a:1) biologists write that the goats were "introduced by sportsmen," although they do not cite references to substantiate their assertion. Early statements do not identify any "sportsmen" involved in the release events, yet Hutchins and Stevens (1981:59) indicate that the release was accomplished "through the combined efforts of Washington state wildlife agents, the U.S. Forest Service, and a local sportsman's club." They do not identify the sportsmen or their club. It seems to be stretching a bit to infer that mountain goats were introduced for the purposes of hunting, particularly in light of Webster's 1920 statement that the Klahhane Club desired an unmolested population of mountain goats.

But the stretch was regularly made in the 1980s, and a factoid was born. Sometimes the assertions employ innuendo, such as Brad Carlquist's (1990:13) statement that "the introduction was a joint project of the local hunter's club and the Clallam County Game Commissioner" and Barbara Sleeper's (1990:18) statement that the goats "were introduced ...by a local sportsmen's club working with Washington State wildlife agents and the U.S. Forest Service." Neither author substantiates the statement with citations of the historical literature. Other assertions are dogmatic: the goats were released "by hopeful trophy hunters in the 1920s" (Luoma 1993:33) as "game animals" (Weingrod 1994:29) and "were meant to flourish for the benefit of hunters" (Wright 1996:23). Curiously, the role of the Forest Service and of state game officials is not mentioned.

The Draft EIS (Olympic National Park 1995:15) states: "The purpose of these goat introductions was *apparently* to establish a population which could be hunted" (emphasis added). Was the issue not resolved by NPS research because it was not seen as germane to the controversy? If so, then why is this factoid repeated so often? If the public is convinced that the mountain goats were originally introduced to be shot by hunters, then that may short-circuit anticipated public outrage at a plan to eradicate the goat population with high-powered rifles. Omitting mention of the government agencies that were also clearly responsible for release of the goats reduces the blame for the goat problem to one source—hunters—and removes any trace of blame from the agency whose task is to eradicate the goats. Ironically, it is the predecessors of the biologists affiliated with the U.S. Biological Survey, who now call for the eradication of the goats (Peterson and Woodward 1993), who originally suggested that mountain goats be released on the Olympic Peninsula.

Although it is true that many early-twentieth-century wildlife-management activities involved the artificial introduction or enhancement of fish and game populations for the benefit of fishers and hunters (Dunlap 1988; Gray 1993), it does not necessarily follow that the introduction of mountain goats to the Olympics was for this reason. Perhaps the goats were introduced to produce an eventually huntable population, although that is not the overall impression I get from the literature. But a poorly researched belief of individuals studying the goat population has become a factoid that readily aligns with the beliefs of those who wish to see the goats eradicated.

EMOTIONS

Victor Scheffer (1993a:918), an early Olympic National Park biologist, proposes that the Fund for Animals personnel, who argue the goats ought to be left alone, should be "relying purely on moral persuasion" and sentiment or emotion. But Scheffer uses scientific—historical, zoological, anthropological, and archaeological evidence to reach his conclusion that goats are not native to the Olympics, as do Olympic National Park personnel (see chapter 6). Thus, Scheffer attempts to disarm Fund for Animals personnel, who also base some of their arguments on scientific evidence (Anunsen and Anunsen 1993). Scheffer (1993a:918) implies that wildlife management should be founded on "a set of preferences *compounded* of sentiment (or emotion) *and* logic (or reason)" (emphasis added). One has to wonder why the Fund for Animals must be restricted to the former when all Olympic National Park documents rely on the latter. As Michael Hutchins (1995:1326) notes in a similar discussion, "we need better ways of resolving disagreements in order to move on to the important work of conservation."

Scheffer (1993a:918), who sides with Olympic National Park in the controversy, suggests that national-park managers "will increasingly become more sensitive to public opinion" (see Carlquist 1990 for a review of Olympic National Park efforts to obtain public input). Scheffer mentions but does not explore a potentially emotional issue: when captured for transplanting, park mountain goats experienced an 8.7% mortality rate in 1988 and a 19% mortality rate in 1989 (Houston, Moorhead, and Olson 1991), a problem known to occur well before these capture events (e.g., Hutchins and Geist 1987). Who could fail to be moved by the fact that some twenty-eight goats were probably literally scared to death as a result of capture myopathy or by the fact that "66% of the adult females captured were lactating, and we suspect that few of their kids, if any, survived" (Houston, Moorhead, and Olson 1991:215)? Scheffer does not mention that sixty-seven kids probably starved to death as a result of the capture of their mothers.

Scheffer suggests that "social scientists" will be increasingly consulted by park managers, but I am unaware of any such consultations in the Olympic goat controversy (Carlquist 1990). Perhaps that is because it is easier to measure the biomass of plants than it is to measure human emotions or opinions (e.g., Johnson et al. 1993). Logic and scientific evidence are to be the basis of park management (USDI–NPS 1988). Science indicates that the goats are not native and their impact on the alpine ecosystem is not "natural." Any kind of logic or scientific evidence to the contrary—indicating that the goats are in fact native—thus invalidates the sole expressed (and mandated) basis for their removal. (I am not saying that emotions should not be considered in wildlife-management decisions; in this I agree with Scheffer.)

The original management plan for the Olympic mountain goats in the 1980s indicates that live capture was to "be used until it was no longer effective or safe" (Carlquist 1990:15; see Olympic National Park 1987a). Olympic National Park staff ranger Paul Crawford states that when the live capture program was discontinued in 1990, "one of the main reasons we cut out live capture was that it just seemed inhumane" (quoted by Luoma 1993:37). This statement concerning the safety of the goats was made in *Wildlife Conservation*, which is not intended to be a scientific journal. In the *Natural Areas Journal*, a scientific journal, biologists responsible for the management of the goats report that the 1990 live capture program was canceled by Superintendent Finnerty because "two independent assessments indicated that continued efforts posed unacceptable risks to the capture team" (Houston, Schreiner, Moorhead, and Olson 1991:89; see Tuler and Janda 1991 and Tuler et al. 1992 for a review of these studies, one of which is Machlis et al. 1990). Here the concern is for the safety of the scientists. In neither journal are both reasons stated.

SCIENCE, HYPOTHESIS TESTING, AND RESEARCH DESIGN

It is difficult to escape the conclusion that the public is being provided with information that is carefully worded to foster sympathy, if not support, for the wildlife managers' goal of eradicating the mountain goats. With the exception of the two hypotheses that Schreiner (1994; Schreiner et al. 1993) tested (see chapter 4), hypotheses have never been explicitly stated in the literature I have examined. A rather large set of variables simply has not been controlled or adequately accounted for because the hypotheses apparently driving the research are poorly conceived. Alternative hypotheses might be derived from questions such as these:

- How does variation in snowpack and the timing and rate of spring runoff influence plant productivity?
- What is the contribution of human urination to the use of Klahhane Ridge by mountain goats?

- What impacts did the early-twentieth-century grazing of cattle and sheep initiate in the alpine ecosystem?
- Do mountain goats exacerbate erosional processes by creating and using wallows?

For example, from the last question we might construct the following hypotheses:

- HYPOTHESIS 1A: Mountain goats exacerbate erosion by using wallows.
- HYPOTHESIS 1B: Mountain goats contribute to erosional processes by creating wallows.

From these two interrelated hypotheses, several alternative hypotheses are immediately apparent:

- ALTERNATIVE HYPOTHESIS 2A: Other vertebrate herbivores exacerbate erosion.
- ALTERNATIVE HYPOTHESIS 2B: Other vertebrate herbivores create wallows.
- ALTERNATIVE HYPOTHESIS 3A: Geological and climatological processes such as runoff from melting snow, rainfall, and wind exacerbate erosion.
- ALTERNATIVE HYPOTHESIS 3B: Geological and climatological processes such as erosion or periods of aridity reduce vegetation cover, which creates exposed sediment that is subsequently used as a wallow by mountain goats.

The alternative hypotheses cited here are sufficient to illustrate my point: typical scientific procedure in general for the last half-century has involved *falsifying* various alternative hypotheses because it is impossible to *prove* that any single hypothesis is correct. If a hypothesis cannot be shown to be false then chances are that it is correct, but only if alternative hypotheses have been shown to be false. None of the alternative hypotheses listed above has been shown to be false by park researchers; in fact, evidence indicates that there is some degree of truth to certain aspects of them.

The next step in the scientific procedure would be to modify the competing explanations so that they were still more specific:

- modified HYPOTHESIS 1A: Mountain goats are *the only* source of exacerbated erosion.

- modified HYPOTHESIS 1B: Mountain goats are *the only* source of the creation of wallows.

While presently available data suggest that these modified hypotheses are probably incorrect, they are treated by park researchers as if they have been proven correct. That is sloppy science. Perhaps the flaws in the research and research design of the park scientists are due to the fact that research has been crisis driven and not well planned and logically designed.

In an analysis of how Olympic National Park personnel planned their resource-management activities prior to the mid-1990s, Peterson et al. (1994: 730) provide some interesting insights:

> The existing planning process at the Olympic NP is not highly structured. As one member of the staff put it, they use the "BOGSAT (bunch of guys/gals sitting around a table) method of planning." In other words, the management staff compiles a wide range of topics, discusses them, prioritizes them, and develops the RMP [resource management plan] with minimal quantitative evaluation and without formal decision-making tools. The result is a large and rather cumbersome document [of over 700 pages].
>
> Despite its shortcomings, the RMP provides the park with a common reference applicable at all levels of park and regional planning. . . .
>
> Most members of the Olympic NP staff indicated that the RMP needed to be upgraded from a reference document to an action plan. Of particular value would be specific information that would facilitate implementation of projects in the field and improve coordination between the main park office and district personnel at various locations. Projects must be written so that field personnel can implement them. . . .
>
> The RMP provides a formal goal-setting process for national parks. . . . but it is also a source of frustration for park personnel. There is nearly always a huge gap between the management programs described in the RMP and the actual programs that are constrained by budget and personnel limitations. Park managers see many critical needs for information, but they also realize that many of these needs will never be filled.

In short, park managers and scientists know what they need to do, but they simply cannot do it due to the way the management plan is constructed and budget limitations. Most importantly, how particular field research projects are to be implemented is not included in the planning document. No wonder there is no set of explicitly stated hypotheses and no long-term research on particular issues. Knowing something about the planning process thus increases our understanding of why the research results are flawed, but it certainly does not excuse the fallacies of those results.

The evidence discussed here and in the preceding two chapters seems to provide a weaker basis for the conclusion that the mountain goats must be eliminated than park biologists would like us to believe, but that evidence does seem to indicate that there *may* be an ecological crisis of unclear magnitude in the Olympics and to hint that part of the cause of that crisis may be the mountain goats. Prior to the total and final eradication of the remaining mountain goat population in the Olympics, however, I believe that all of the available research and data should be thoroughly reviewed by an objective panel of experts. That panel should be given sufficient time and resources to review all relevant primary literature, not just park-produced synopses.

6

HISTORY, ETHNOGRAPHY, AND ARCHAEOLOGY

The heart of the Olympic mountain-goat controversy resides in the park biologists' belief that mountain goats were never present in the Olympics prior to their 1925 introduction by humans (Moorhead and Stevens 1982). Others believe that goats may have been there prior to 1925 (Anunsen and Anunsen 1993; Lyman 1988a). This has been referred to as the "native versus exotic" issue.

EARLY STUDIES

The first formal examination of the native versus exotic issue was published by Moorhead and Stevens (1982). Their discussion appears in modified form in the first unpublished report on the goats produced by the park (Olympic National Park 1981). Moorhead and Stevens (1982:46) begin with this statement: "Although native to the Cascade Range of central Washington, the mountain goat, *Oreamnos americanus,* was introduced to the Olympic Mountains of northwestern Washington by man prior to creation of Olympic National Park. There is no historical evidence of goats in the Olympic Mountains before release of 11 or 12 from Canada and Alaska between 1925 and 1929."

Moorhead and Stevens (1982:47, 48) also indicate that (1) "the fossil record of mountain goats is scant"; (2) "no early [zooarchaeological] faunal remains or ethnographic evidence of mountain goats have been reported"; and (3) early historic reports by zoologists who explored the area indicate that they did not find mountain goats. They fail to note that the vertebrate paleontological research that has been done on the Olympic Peninsula involves materials that are millions of years old (Barnes and Goedert 1996 and references therein) and thus is irrelevant to demonstrating the presence or absence of native goats. Virtually no paleontological research on vertebrate remains dating to the last 10,000 years has been done, so it should come as no surprise that no "fossil" evidence of mountain goats is available.

Moorhead and Stevens (1982:49) apparently reasoned that because there was no prehistoric evidence of native goats, mountain goats must never have been there. A fossil typically is defined as some remaining trace of a long-dead organism, such as a bone, tooth, seed, or footprint (the last is referred to as a "trace fossil"). Seldom mentioned but implicitly understood by paleontologists and most zooarchaeologists is the fact that "fossils" are at least 10,000 years old; if a bone or seed is younger than that, it is often referred to as a "subfossil."

Moorhead and Stevens (1982:47) apparently consulted two anthropologists rather than reading the relevant literature to determine if ethnographic or archaeological evidence of goats existed. They do not mention the prehistoric caribou (*Rangifer tarandus*) remains found on the Olympic Peninsula (see below), even though they talked with the individual who found them. Nor do they mention that the majority of the zooarchaeological evidence came from the far northwestern corner of the Olympic Peninsula and consisted largely of seal and sea-lion remains. That evidence suggests that the prehistoric people who collected and deposited those bones and teeth were exploiting sea mammals but perhaps not alpine mammals and thus may be inappropriate evidence for concluding that mountain goats were never present in the Olympics.

Moorhead and Stevens (1982:47) refer to Scheffer's (1946:13–15) list of early scientific collecting expeditions and note that he found no one who reported collecting mountain goats. They also state that in his overview of Washington mammals Walter Dalquest (1948) "makes no mention of goats in the Olympics" (Moorhead and Stevens 1982:47). This statement is paradoxical because goats were clearly present there at the time Dalquest wrote. Dalquest (1948:110) indicates that the "principal purpose of this report is to describe and interpret the distribution of the various species and races of mammals that occur in Washington." Thus Dalquest (1948:61–63) discusses a conjectural biogeographic history of Olympic Peninsula mammals and explicitly omits consideration of taxa introduced or exotic to the state (Dalquest 1948:111).

It would have made more sense for Moorhead and Stevens (1982) to note that Ernest Booth (1947:598, 602, 603), who shared data with Dalquest, reports that mountain goats were introduced "from Alaska and Banff, Alberta, into the Olympic Mountains." They do not reference his work, however. Nor do they refer to an early mention of the goats on the peninsula in the Works Progress Administration–sponsored Washington Writers' Program (1950). The (anonymous) authors list many of the mammals found in Olympic National Park: the Roosevelt elk is "native to the Olympic Peninsula....The cougar, Columbia black-tailed deer and Rocky Mountain mule deer, the mountain goat, black bear, Olympic mountain lion, and northwestern wildcat also live in the park" (Washington Writers' Program 1950:603). No mention is made of the fact that the mountain goats might be exotic and that goats were introduced to the Olympic Peninsula in the 1920s.

Moorhead and Stevens (1982) apparently relied on Scheffer's (1946) report, adding only the citation of Dalquest (1948). Scheffer (1946:13), in turn, relied mostly on "information from a paper by" E. S. Hall (1932). According to Hall's (1932) history of several of the expeditions listed by Scheffer, many of these early scientists never got close to the modern range of mountain goats (Lyman 1988a:14). If these scientists were never in goat territory and if the 1850–1900 range of mountain goats approximated their range in the 1980s, then it is no surprise that they never saw any mountain goats.

Moorhead and Stevens's (1982) research was published *after* the "experimental" goat-removal program began in 1981 (Olympic National Park 1981, 1982), though their results were known as early as 1979. Their research was not thorough, as we shall see, and yet it was the warrant for the goat-removal program during the 1980s. While potential weaknesses in Moorhead and Stevens's (1982) research were pointed out in 1988 (Lyman 1988a), park researchers did not attempt to improve on that research until early 1992, when the existence of evidence potentially contradicting their conclusion was publicized.

In addition, some potentially contradictory evidence was known to Moorhead in the mid-1980s, yet he chose to ignore it. In July 1983 he received a copy of S. C. Gilman's (1896) report, which was not cited by Scheffer (1946) or by Moorhead and Stevens (1982), from someone identified in park files only as "Carl." Moorhead's statement that "there is no prior record of mountain goats in the Olympics" (1988a:10) ignores—rather than attempts to discredit—Gilman's contradictory evidence suggesting that goats were present in the Olympic Mountains in the 1890s. Park researchers have recently admitted that "it would have been more open and prudent if the early observations had been mentioned in [Olympic National Park 1987a] and the rationale given for not considering them valid records" (reported in Wagner et al. 1995:105). By not acknowledging contradictory evidence, park researchers have left themselves open to criticism (e.g., Anunsen and Anunsen 1993; Lyman 1994, 1995a).

NEGATIVE EVIDENCE

As historian D. H. Fischer (1970:62) notes, "The nonexistence of an object is established not by nonexistent [negative] evidence but by affirmative evidence of the fact that it did not, or could not exist." "The fallacy of negative proof" occurs "whenever a [researcher] declares that 'there is no evidence that *X* is the case,' and then proceeds to affirm or assume that not-*X* is the case" (Fischer 1970:47). Fischer (1970:47) argues that "the only correct empirical procedure is to find affirmative evidence of not-*X*—which is often difficult, but never in my experience impossible." Moorhead and Stevens (1982) offered arguments that goats were exotic on the basis of negative evidence that were vulnerable to the problem of Fischer's negative proof. What about potential affirmative evidence?

In his account of a 1907 ascent of Mount Olympus, Belmore Browne (1910:887) reports that no evidence of mountain goats was found and notes that "it is unquestionably true that if Rocky Mountain goats were liberated in the Olympic Mountains they would thrive and would be a valuable addition to the wildlife of the region." Because they spent very few days in the mountains, I wonder how thoroughly they examined the terrain for wildlife. Browne's ascent is not mentioned by Schultz (1993, 1994), despite the fact that she was informed of this event prior to publication of her report (Lyman letter of June 21, 1993, to Olympic National Park). Schultz (1993, 1994:278) mentions that "The Mountaineers" of Seattle also climbed Mount Olympus in 1907 and in later years and that none of the published materials regarding these expeditions "provided extensive wildlife observations." She cites none of these references (e.g., Curtis 1908; Lee 1913), however, so it is difficult to determine if the evidence therein should be considered negative or not.

In his discussion (cited by park scientists) of the mammals he collected from the Olympic Mountains, D. G. Elliot (1899:244) reports that while camped in the dense coniferous forests, "I do not think I was ever before in a country that was apparently so devoid of animal life." Later, while crossing ridges between the canyons of major rivers, Elliot

(1899:246) notes that "great fires were now raging in the mountains, as no rain had fallen for six weeks, and dense smoke obscured all the view and effectually put an end to photographing." Not only was the set of mammals collected by Elliot incomplete (he collected specimens of twenty-eight of the forty-one mammalian species today known to occur in the Olympics), but the sightability of mammals was seriously impaired. Thus while Browne's and Elliot's reports might be taken as affirmative evidence for the absence of native mountain goats, it is weak affirmative evidence at best. Neither the fact that Elliot's collection was incomplete nor the fact that his visibility was impaired is mentioned by park researchers.

Efforts to establish the twentieth-century presence of mammals on various western mountains are still underway (e.g., Grayson and Livingston 1993), underscoring what we do *not* know about modern western mammalian biogeography as well as the inaccuracy of the late-nineteenth- and early-twentieth-century historical record. Stewart Brandborg (1955:15–16) made this point forty years ago with regard to establishing the historic range of mountain goats in Idaho. For park researchers to argue that early naturalists such as Elliot (1899) are to be trusted because they collected several hundred specimens of mammals, " 'with few exceptions embracing all the species known to inhabit the region' (Elliot 1899:246)" (Schultz 1994:284), is to clutch at straws. Elliot had no idea of what he should or should not find; that is why he was there in the first place. A complete inventory of the mammals of the Olympic Peninsula was not available until shortly before Scheffer completed his 1946 report (Lyman 1994). Prior to the completion of Scheffer's report, it was never clear which species of mammals were present in the Olympics.

Using E. Raymond Hall's (1981) taxonomy, including only nonvolant, terrestrial mammals and excluding the coyote (*Canis latrans*) and porcupine as they are apparently recent immigrants to the Olympic Peninsula (see the following references), Elliot (1899) listed twenty-eight species, Albert Reagan (1909) thirty species, E. B. Webster (1920) thirty-seven species, Walter Taylor and William Shaw (1929) thirty-five species, Lee Dice (1932) twenty-four species, and Arthur Svihla and Ruth Svihla (1933) thirty-seven species for the peninsula. Only Scheffer (1946) and Murray Johnson and Sherry Johnson (1952) list the forty-one species historically known to occur on the Olympic Peninsula. While taxa missing from the pre-1935 species lists were found by biologists and various explorers during their surveys of the Olympics, their reports were typically not published and thus are not well known. Only by consulting the species lists provided by, minimally, Webster (1920) and Taylor and Shaw (1929) can all forty-one species be found (even then, there are errors in the list).

LATER STUDIES: HISTORICAL EVIDENCE

Olympic National Park historian Susan Schultz (1993, 1994) offers a relatively thorough review of much—but not all—of the historical evidence bearing on the pre-1925 presence or absence of mountain goats in the Olympics. Her goal is to "provide a thorough, careful, and balanced consideration of all located records that pertain to the issue" of whether mountain goats were present in the Olympics prior to 1925 (Schultz 1994:257). In my view, she fails on all counts.

Early Knowledge of the Range of Mountain Goats

Schultz (1994:287) states that "early naturalists describing the known range of mountain goats in North America do not include the Olympics (e.g., Fannin 1890; Hornaday 1904; Seton 1909)." In fact, Fannin, curator of the Provincial Museum in Victoria, British Columbia, coauthored an article that indicates that mountain goat range *did* (at least prior to 1925) include the Olympics (Fannin and Grinnell 1890c). In the article cited by Schultz, Fannin (1890:343–44) states that "there are few animals of the North American Continent of which, having regard to its distribution and relative abundance, so little is known as of the Rocky Mountain Goat." Fannin (1890:343) notes the southern (36°N latitude) and northern (62°N latitude) limits of the known distribution of this species and that it ranges from the British Columbia "coast-line to the Rockies" and is "probably most abundant along the rugged peaks of the Coast Range." He does not state where the goat is not found. For Schultz to imply that Fannin does not include the Olympics in the goats' range because they were not present there badly misrepresents what he said. He explicitly indicated that the range of this species was very poorly known, a fact overlooked by Schultz.

In one of the earliest statements I have found describing the geographic distribution of mountain goats, Hallock (1879:40) writes: "The White Goat is confined to the loftiest peaks of the Rocky Mountains: it is not known south of Colorado, and is probably rare south of Washington

Territory, but is found to the northward as far as Alaska." Twenty-five years later, we find Owen Wister's (1904:249–50) lament:

> It is surprising, indeed, that at this late day, when investigation and verification are so easy, no naturalist seems anywhere to have written a plain, complete paragraph answering the plain, natural question: In what states and territories does the white goat live? It would seem the naturalist's business to tell us this. We have the right to expect to open some single standard book, and find such facts at once. Well, I have had to open eight, gathering here a fact and there a fact....the following is the most detailed information to be found so far set down in any one place.
>
> In Alaska and British Columbia we find the goat, and in northwest Montana, and in Idaho, but only in spots; he is also in the northern Cascades in Washington, but, oddly enough it appears, not in the Olympic Range. Nor is he in the southern Cascades, in Oregon. Elsewhere he is not, unless possibly in California.

While Wister's (1904) account is one of the most detailed to that time, it does not indicate exactly where in British Columbia, Washington, and California the goat is to be found. Is Wister correct and are Fannin and Grinnell (1890c) incorrect? Unfortunately, none of these authors cite references. Wister states that it only "appears" that the mountain goat is not found in the Olympics. Clearly, his information is second-hand and thus less than solid.

In another reference not cited by Schultz (1993, 1994), Dwight Huntington (1904) discusses the range of the mountain goat:

> It is nowhere abundant in the main chain of the Rocky Mountains, if, indeed, it exists in these mountains. Owen Wister tells us we may expect to find white goats as far south as the Saw Tooth Mountains, Idaho, and among the peaks northward from Lake Chelan, the Okanogan and Methow rivers, all three in Washington, and also on many mountains near the coast in British Columbia....The best places to find the white goats are in the mountains reached by the Great Northern, the Northern and Canadian Pacific Railways. The surveying parties reported them from many places in the Cascade Mountains last year....

> Some writers say we may find a few of these curious animals in isolated colonies in the high mountains in Wyoming, Colorado and New Mexico. Owen Wister, however, gives the range as extending from Alaska to Montana and Idaho, but only in "spots" in the two States named. Goats are also found, he says, in the northern Cascades in Washington, but not in the Olympic range nor in the southern Cascades in Oregon. (Huntington 1904:218, 219, 227)

Huntington's report is important because it indicates that the natural history information of many writers around the turn of the century was secondhand. Huntington cites Wister (1904), who, as we have seen, had to examine "eight" references to compile the information. Interestingly, Wister apparently "did not believe until 1889 there was such a thing as this goat" (Huntington 1904:219).

Madison Grant (1905:236–37), whose report Schultz (1994:281) cites with favor, reports that goats were present on Mount Jefferson of the central Oregon Cascade Mountains at the turn of the century and that they previously were found on Mount Hood in the Cascades of northern Oregon. No other biologist has ever agreed with this assessment of the range of mountain goats, with the exception of Hall (1981:1111–12), who lists them as occurring on Mount Jefferson but cites no references for this and does not list Grant (1905) in his references. The Mount Jefferson record is attributed to Grant (1905) in the first edition of Hall's book (Hall and Kelson 1959:1027). In what seems to be a direct contradiction of Grant's report, Vernon Bailey (1931:156), whose byline reads "Chief Field Naturalist of the Biological Survey," advocates transplanting surplus goats from Mount Rainier to the higher peaks in the Oregon Cascades (Hood, Jefferson, Three Sisters, and others) in order that this "attractive form of native life be given a wider range." Thus, even in the early twentieth century our knowledge of the distribution of mountain goats was contradictory and incomplete.

Mountain Goats in the Olympics?

Schultz (1993, 1994) cites two articles on mountain goat behavior and ecology by John Fannin and George Bird Grinnell (1890a, 1890b) published in the weekly periodical *Forest and Stream.* She does not cite their article "Range of the White Goat," published in that same journal (Fannin and Grinnell 1890c).

Fannin and Grinnell (1890c) describe several specific locations in western North America where mountain goats were found in the late nineteenth century:

> In the State of Washington [mountain goats] are *found in considerable numbers on the high mountains to the west of Puget Sound*, and on Mount Rainier. Mr. L. L. Bales writes us that they are common on Mt. Baker, in Whatcom county, and thence northward. They are also abundant on the *Olympian Range mountains through which flow the Quinneault [Quinault], Quillinot, Duckeybush [Duckabush], Docewallup [Dosewallips], Dungeness and Skokomish rivers*, on the Smilkameen and its tributaries, and Mt. Shupaka [Chopaka], and generally throughout the state, wherever there are mountains adapted to them. (Fannin and Grinnell 1890c:62; emphasis added)

One might wonder if the first indication that goats occurred in the Olympics in the 1880s was merely a typographical error. The list of rivers flowing from the "Olympian Range" clearly identifies what are today known as the Olympic Mountains, however. Gilman (1896) may have been using Barnes's (1890) report to conclude that mountain goats were present in the Olympics, but Fannin and Grinnell (1890c) could not have been using it because their article was published in mid-February of 1890, well before his report. Barnes's (1890) report was not based on Fannin and Grinnell's (see below).

Forest and Stream, which has been called "the most influential outdoor journal of its day," was founded in 1873 by Charles Hallock, a "distinguished conservationist" (Merritt 1995:42). G. B. Grinnell became natural-history editor in 1876, having served as an advisor since the journal's inception (Reiger 1992); he took over as owner and editor in 1880. Grinnell was present at the dinner party in December 1887 when the Boone and Crockett Club was formed; also present at that dinner was his friend Madison Grant. In a paper published early in the twentieth century, Grant (1905:236), president of the New York Zoological Society (Twight 1983:46), states that the mountain goat "is absent from the Olympic Mountains."

How could Grinnell and Grant, both interested in conservation issues at Yellowstone National Park (Reiger 1992), come to such different conclusions regarding the presence or absence of mountain goats in the Olympics? Clearly, they corresponded with one another (Reiger 1992). Is there perhaps an answer to the question in their as-yet-unstudied correspondence?

Apparently, Schultz did not discover Fannin and Grinnell's (1890c) report and thus did not include it in her review (Houston 1995). Both were reputable naturalists, and both are cited with favor by Schultz (1993, 1994). Their report cannot be discounted as resulting from a misidentification of the species; Fannin and Grinnell (1890c) explicitly distinguish between mountain or bighorn sheep and mountain goats, although they use taxonomic names that have since been discarded (*Ovis montana* and *Mazama montana*, respectively). But park scientists discount this report because "there is a mountain of evidence to the contrary" and because Fannin and Grinnell "mistakenly reported the range of mountain goat included Colorado [and] they place Mount Saint Helens and Mount Adams in Oregon instead of Washington" (Houston 1995:239). Washington became a state in 1889, and perhaps this is why they erred; their map shows no state lines. Further, it seems that the range of mountain goats did once include Colorado (Irby and Chappell 1994).

If we discount everyone in the historical literature who made an error, we would have no literature left and the "mountain of evidence to the contrary" would disappear. For example, C. Hart Merriam, first director of the U.S. Biological Survey (later to become the U.S. Fish and Wildlife Service), spent two weeks in the Puget Sound area and five days at Neah Bay in 1888 (Schultz 1993:38–41; 1994:282–84). He writes that several mammalian species were "said not to occur in the Olympics [including] sheep, goats, porcupines, coyotes, foxes, [and] grisly bear" (quoted in Schultz 1993:40; 1994:283). In noting that goats are "said" not to occur in the Olympics, Merriam is apparently using secondhand information, a point not acknowledged by park biologists or by Schultz (1993, 1994).

Schultz (1993, 1994) values Merriam's views highly. But we now know that Merriam was wrong on at least one count; the Olympics do contain one mammalian species unique to that area—the Olympic marmot. This is based on the late-twentieth-century conception of mammalian taxonomy, and thus perhaps we cannot fault Merriam for his statement. But he made other mistakes. In his unpublished 1888 field journal Merriam writes that "there are no porcupines, muskrats, or rats near Neah Bay, and the Indians do not know any such animals. Deer and elk occur a few miles back in the deep forests. Caribou are not known on this side of the Strait" (quoted in Schultz 1993:39; 1994:282). In an unpublished manuscript Merriam writes: "The fauna [of the Olympics], at least as far as the land vertebrates are concerned, is rather meager. ...Among the mammals, perhaps the most striking peculiarity of the region is the species it lacks. There are no mountain sheep, goats, porcupines, coyotes, foxes, or grizzly bears in the Olympics" (quoted in Schultz 1993:41; 1994:283).

Muskrats (*Ondatra zibethicus*) were unknown to Indians living in the northwestern corner of the Olympic Peninsula and Merriam apparently failed to collect any, so he concluded that this small mammal did not occur there (cited in Schultz 1993:39; 1994:282). Subsequent research by other zoologists (e.g., Johnson and Johnson 1952; Scheffer 1946) revealed the twentieth-century presence of this species, and archaeological research (Huelsbeck 1983, 1994) revealed the presence of this species over the last 1,000 years in precisely the geographic location where Merriam worked. Schultz fails to note this natural-history error. If we are to discount Fannin and Grinnell (1890c) because they "got it wrong" (Houston 1995:239) concerning details of natural history, then we must discount Merriam because he too got details wrong.

Schultz's research is not thorough and is noticeably biased. For example, she visited the Bancroft Library to inspect the papers of C. Hart Merriam. In those papers she found "a popular article on the mountain goat authored by Merriam," but the periodical or journal in which it appeared was not identified so Schultz could not provide a specific reference. I identified the unknown periodical as *Country Life in America* (Merriam 1902) by inspecting obituary and biographical sketches of Merriam (Grinnell 1943; Osgood 1943, 1947).

Schultz also found "an article by [Merriam] on the taxonomy of the mountain goat (Merriam 1895) *showing Merriam's particular interest in this animal*" (Schultz 1994:283; emphasis added). This "particular interest" is not evident in his publications. Merriam published over 600 books, monographs, articles, and book reviews between 1873 and 1934, inclusively (see Grinnell 1943 for a nearly complete list). Of these, only the two mentioned by Schultz plus a third she did not mention (Merriam 1897b) explicitly concern the mountain goat. During his career, Merriam described and named over 660 new species and subspecies of mammals as part of his job with the Biological Survey. Such naming was typical of early North American biologists (Brown and Wilson 1994; Engstrom et al. 1994), who believed that every distinct form of mammal should be described as a new species (e.g., Merriam 1897a, 1919). It is fallacious to suppose that one article on the taxonomy of mountain goats signifies a "particular interest" in this species on Merriam's part. It would be far more convincing to argue that he had a particular interest in anthropology (thirty published articles) or in bears (twenty-one titles). These two topics dominated at least the last half of his career (Osgood 1943, 1947; Palmer 1954; Sterling 1989). The two most recently published synopses of mountain-goat biology (Rideout and Hoffmann 1975; Wigal and Coggins 1982) do not cite any of Merriam's papers on goats; nor does Chadwick (1983) cite his work.

Schultz's bias permeates her evaluation of the historical literature concerning the presence or absence of mountain goats in the Olympics.

Uneven Treatment: Reports of Goats

Schultz (1993, 1994) reproduces the following quote from Samuel C. Gilman's (1896:138) report: "Game is plentiful, and it would be a paradise for the hunter were it not so difficult of access. In addition to elk and bear, before mentioned, are deer, mountain goat, cougar, beaver, otter, fisher, wildcat, marmot, geese, ducks, grouse, partridge, quail, pelican, and many smaller or less desirable birds and animals." Schultz (1993:20, 22, 51; 1994:270, 290) tells us that this statement was made by "a civil engineer rather than a naturalist" and "is not related to any specific observation of the animal." Gilman died prior to acceptance of the article for publication. According to Schultz (1993:21, 52; 1994:271, 290), "this leaves open the possibility that the article may have been considerably edited and altered, and that Gilman would not have been able to correct any inaccuracies," and that the article "may have been edited without his review." Schultz (1993:22; 1994:271) also notes that partridge and quail are not known to occur naturally on the peninsula, though quail are an "introduced species." In making these points, she apparently wishes to destroy Gilman's credibility as a naturalist.

Schultz's suggestion that Gilman's article indicates that mountain goats were native to the Olympic Peninsula only because of editorial whim is contrived. Why should the editors of *National Geographic* make such a change? The fact that Gilman did not explicitly state that he "personally" had observed goats does not in my view weaken the significance of his statement. He did not explicitly state that he had "personally observed" many of the animals he lists. Maybe he simply included mention of the mountain goat because he had read Barnes's (1890) report; Gilman was, after all, well aware of the *Press* Expedition ("In the Olympic Mountains" 1890). Schultz does not mention that Gilman was quite familiar with the territory (he apparently made three separate trips into the mountains).

Schultz (1993:22; 1994:272–73) quotes the newspaper story authored by C. A. Barnes (1890:20) on the *Press* Expedition:

> The mountains are alive with elk, for the most time very tame. Some bands, however, having been out upon the foothills, and having probably been chased, are more wild. Deer are also plentiful. One goat was seen by the party. Grouse, pheasant and chicken are undoubtedly plentiful, although we saw few, owing to the severity of the season. Beaver and fisher are numerous on the Quinault. Black bear are plentiful during the proper season. We saw one track of a cinamon [*sic*] bear on the Goldie river. The cougar, or mountain lion, are numerous. We frequently saw their tracks, but never the animal himself. The grey wolf and the wildcat are common.

Schultz (1993:23; 1994:273) then writes:

> Barnes does not specify that it was a mountain goat, and, although he is presumably summarizing wildlife found in the Olympics, it is possible the explorers saw a free-ranging goat belonging to a settler in the Elwha or the Quinault Valley. He does not mention where or when or under what circumstances the goat was sighted. Nor is the observation of a goat recorded in his or Christie's diary narrative. The mention of "chicken" in the next sentence gives cause to wonder whether Barnes is including domestic animals; there are no prairie chickens in the Olympics. Neither are there "cinnamon" bears in the Olympics. Probably Barnes thought the track may have belonged to a grizzly bear because of its size. [In the 1994 report, the last two sentences read:] Although "cinnamon" is today considered one of the brown color phases of the black bear, Barnes may have thought the bear track he mentioned belonged to a grizzly bear because of its size. There are only black bear on the Olympic Peninsula, and they all exhibit a black color phase.

Schultz is unsure if Barnes was actually recording "wildlife" rather than just animals the expedition members saw and Barnes is untrustworthy as a naturalist because he did not know the difference between black bears (*Ursus americanus*) and grizzly bears (*Ursus arctos*). Again, the credibility of the witness as a naturalist is called into question.

Schultz makes several unwarranted but implicit assumptions in her assault on the credibility of Barnes and Gilman. Perhaps the most subtle is that the alignment of folk taxonomy (common names of animals) with the Linnaean taxonomy (scientific names of animals) is the same today as it was a century ago when Barnes and Gilman wrote. Schultz does note that in the late nineteenth century "the term 'cinnamon bear' was apparently synonymous with [the twentieth-century term] grizzly bear" (Schultz 1993:20; this statement is not included in Schultz 1994). But she is the one who synonymizes Barnes's "cinnamon bear" with "grizzly bear," not Barnes.

I believe Schultz makes an error when she equates Gilman's "partridge" and "quail" with *Perdix perdix* (gray partridge) or *Alectoris graeca* (chukar partridge), both introduced to Washington in the twentieth century, and the four species of quail introduced to the state in the twentieth century, respectively (Larrison and Sonnenberg 1968); Gilman probably meant grouse. She also errs, I think, when she equates Barnes's "chicken" and "goat" with *Gallus gallus* (domestic chicken) and *Capra hircus* (domestic goat), respectively. Barnes did not record the countless horses and cows and dogs he saw on the trip, although he did mention some domestic animals; if he had recorded a stray domestic chicken or a stray or feral domestic goat in the Olympic Mountains, I suspect that he would have commented on the fact. Finally, Barnes indicated that they saw few "chickens" because it was winter. This suggests that he meant grouse.

Schultz seems to have lost sight of the reason that Barnes was in the Olympics: he and his comrades were exploring an unknown territory. They had no reason *not* to expect to find mountain goats. I find it remarkable that they managed to see a goat at all; their efforts were focused on surviving and getting through the mountains in the middle of winter when game animals were scarce at high altitudes (Majors 1981a, 1981b, 1981c, 1981d, 1984a, 1984b, 1984c; Wood 1967) and "beasts the color of winter" would be difficult to see.

Did Barnes, Christie, and the others really see a mountain goat? "It is well known that the ewe [female] bighorn [sheep] (*Ovis canadensis*) was often called a goat" by early explorers (Brandborg 1955:16; see also Armstrong 1972:315). This was perhaps not a problem in this case, however. James A. Christie, co-leader of the *Press* Expedition, had spent three years "in the arctic region of northwestern Canada. During that time he had explored the Liard, Peace, and Mackenzie rivers" (Wood 1967:17). The headwaters of any of these rivers would have been mountain-goat country. Thus, I find it unlikely that he would mistake a domestic goat for *Oreamnos.* Charles A. Barnes had previously "explored the mountains in Arizona, Oregon, and Washington" (Wood 1967:20). If he explored Mount Rainier—a popular place even then (Haines 1962)—he would probably have seen mountain goats and thus would have known a mountain goat when he saw one. It is not surprising that the expedition report should not make a major point of a single sighting of an animal that might have been familiar to the two leaders. Their daily journals focused on phenomena that bore directly on their survival and had to be dealt with on a day-to-day basis (see Majors 1981a, 1981b, 1981c, 1981d, 1984a, 1984b, 1984c; Wood 1967). Thus, Schultz's (1993:51; 1994:290) suggestion that "if a mountain goat is meant [by Barnes (1890)], it is unusual that the sighting of such a game animal in the Olympics would not have been noted in the detailed daily journal entries of the explorers at the time it occurred" is contrived.

Schultz's evaluation of these two reports of native mountain goats often reads like special pleading. There is no attempt to place these reports into their historic or personal contexts. This does not mean that I think these two reports demonstrate that mountain goats were present in the Olympics in the 1890s. But Schultz seems intent on demolishing the credibility of the reporters and questioning the veracity of the records rather than evaluating those records thoroughly and objectively.

Uneven Treatment: Reports of No Goats

Schultz's treatment of reports and journals written by zoologists, explorers, and residents of the Olympic Peninsula in the late nineteenth and early twentieth centuries that indicate that mountain goats did *not* occur on the peninsula is quite different.

Schultz notes that James Wickersham (1961), who explored the southeastern portion of the Olympics in 1889–90, "did not mention observing mountain goats" in his single known document on that exploration (Schultz 1993:28; 1994:276). She does not mention that Wickersham was not a naturalist and that the purpose of his report was not to document the natural history of the area.

Albert B. Reagan wrote two reports on the natural history of the peninsula and a third that was explicitly concerned with the mammals (Reagan 1908a, 1908b, 1909). Schultz (1993:32–33; 1994:279) states that in those reports Reagan "does not mention mountain goats." She fails to heed Brandborg's (1955:16) caution that the "absence of references to mountain goats in early journals is not proof that they did not exist in an area. The route of travelers along valley bottoms and through open terrain during mid-summer, when the goats were at high elevations, precluded observation of them." We do not know where Reagan went to compile the data for his studies. Schultz also fails to note that Reagan was not a naturalist and that his list of Olympic Peninsula mammals included only thirty of the forty-one terrestrial, nonvolant mammalian species presently known to occur there. She quotes Reagan's (1908b:19; 1909:137) statement that "pheasants" were year-round residents of the peninsula. If Reagan meant "grouse," he was correct; no birds today termed pheasants are native to the peninsula. Others also mention the presence of pheasants "in all parts of the [Olympic] region" (Peters 1936:170), which leads me to suspect that around the turn of the century this name was applied to what are today termed grouse. But Schultz does not evaluate the folk taxonomy in reports that do not contradict her belief that goats were never present in the Olympics prior to 1925.

Schultz (1993:33–34; 1994:279–80) states that in his autobiography Chris Morgenroth (1991), "one of the first forest rangers on the Olympic Peninsula," who worked on the forest reserve and the national forest between 1903 and 1927, "never mentions the presence of mountain goats in the Olympics before their [1925] introduction." In fact, Morgenroth never mentions mountain goats in his autobiography at all, even in entries made after 1925. Rather, the single mention of this species in his book (which was compiled and edited by his daughter) is found in a paragraph authored in 1946 by Clarence Adams, an administrative assistant with the U.S. Forest Service: "In the summer of 1923 ... Chris [Morgenroth] and I turned two pair of mountain goats loose near Storm King Ranger Station" (in Morgenroth 1991:166). This may be a typographical or editorial error. Adams indicates not only the wrong year, but the wrong season. The goats were released on January 1, 1925 ("Introduction of Mountain Goats" 1925). Schultz (1993:34; 1994:280) notes that Morgenroth (1991:48) reports that "pheasants" and "quail" are present in the Olympics, but she does not evaluate his folk taxonomy or mention that he was not a naturalist.

In a summary report on the Olympic National Forest quoted by Schultz (1993:34; 1994:280), Morgenroth writes that "it would be advisable to stock the forest with a few mountain *sheep*[;] they would undoubtedly thrive in the higher altitudes" (emphasis added). Schultz (1993: 34; 1994:280) then states that "it is unlikely [Morgenroth] would have made such a comment if mountain *goats* already inhabited the Olympic Mountains" (emphasis added). Certainly she does not mean to imply that mountain sheep and mountain goats are the same species. While the ranges and habitats of these two ungulates overlap, they are not identical (e.g., Chadwick 1983; Laundré 1994). Why then would it be "unlikely" that Morgenroth would make such a suggestion?

Schultz (1993:34–35; 1994:280) notes that Edward B. Webster's 1920 book *The King of the Olympics: The Roosevelt Elk and Other Mammals of the Olympic Mountains* "won high praise from naturalists, including C. Hart Merriam, former chief of the U.S. Biological Survey." Webster (1920:146) writes that "the fauna of the Olympic Mountains has, *for some peculiar reason*, never included the mountain goat" (emphasis added). The emphasized phrase is not included in the quotation given by Schultz (1993:35; 1994:280); she indicates that the quotation is a "positive" statement that goats have never been present in the Olympics. That Webster found it "peculiar" suggests to me that he might not have been as positive as she would have us believe. Schultz does not mention that Webster was not a naturalist. His book includes a discussion of the cottontail rabbit (*Sylvilagus* sp.), a genus known today not to occur naturally on the peninsula. Schultz does not mention this natural history error.

Schultz (1993:42; 1994:284) notes that the 1898 expedition of the Field Museum of Natural History (Chicago) camped at an elevation of about 4,300 feet in areas that are "good mountain goat habitat." She does not note that this was the first report on Olympic Peninsula mammals to be written and published by a professional zoologist and that Elliot (1899) described only twenty-eight of the forty-one species today known to occur there. Elliot made no remarks about which animals he expected to find but failed to find, so the fact that he did not mention the mountain goat does not mean that goats were not present, as Schultz implies.

Schultz (1993, 1994) does not quote Elliot's (1899:246) remark that at one point during the expedition "great fires were now raging in the mountains, as no rain had fallen for six weeks, and dense smoke obscured all the view and effectually put an end to photographing." Maybe this is why Elliot reported no mountain goats. Nor does Schultz note that the expedition only spent about half the time in (modern) mountain-goat range.

During their expedition up the Quinault River in the summer of 1890, the Gilmans also encountered a visibility problem. They noted that "smoke of burning timber interfered with the view from the mountain tops" ("In the Olympic Mountains" 1890). Early settlers noted that "prior to 1915, forest fires were a common occurrence in the summer season and dense clouds of smoke were expected to hide the scenery and obscure the sky" (Morgenroth 1991:115). Ruby El Hult (1954:115) makes a virtually identical statement. Schultz does not mention the clear implications of these observations for the possibility of spotting mountain goats.

Uneven Treatment: Discussion

Schultz (1993:46; 1994:287) concludes that

> at least four pre-1925 discussions of Olympic Peninsula wildlife state positively that goats were not present. Three of these authors had explored the Olympic Mountains. Early naturalists describing the known range of mountain goats in North America do not include the Olympics. Several biologists writing about Washington mammals after the 1920s state that mountain goats were introduced to the Olympic Mountains. The remainder of the reports by recreationists and wildlife biologists that list Olympic mammals do not include goats.

Schultz (1994:287) indicates that the positive statements that goats are not present include those by Webster (1920), Merriam (unpublished manuscript cited by Schultz), Grant (1905), who had not explored the Olympics, and Private H. Fisher (real name James Hanmore), who was a member of Lieutenant O'Neil's 1890 expedition. The authors of the first two statements made natural-history errors, and Grant's report of goats in Oregon is based on secondhand information and typically thought to be in error. Further, Webster was not a naturalist by training. Applying these criteria equally to those who report that goats were present and those who report that goats were not present, the reports of Webster, Grant, and Merriam can be discounted. That leaves Fisher.

Fisher was not a trained naturalist and had been dishonorably discharged from the service (as Schultz indicates). Schultz (1993:26–27, 1994:275) notes that Fisher used a telescope (of unknown power and resolution) and apparently kept a vigilant lookout for mountain goats, but she fails to mention that the members of the *Press* Expedition also had a "field glass" that they used to search the countryside for game animals. Schultz (1993:28; 1994:276) indicates that "it seems likely, therefore, that if goats had been present, Fisher . . . with his telescope over a period of three months, would have spotted some," especially in light of the fact that the governor of the territory had said that mountain goats are gregarious "and readily noticed in the distance" (quoted in Schultz 1993:28; 1994:276). Such a statement assumes that all goats occur in herds and that Fisher spent his three months in the Olympics in mountain-goat range. Neither assumption is correct. In spite of all this, Schultz (1993:52; 1994:290) takes Fisher's narrative as affirmative evidence that goats were not present in the Olympic Mountains in the early 1890s.

That several biologists writing after 1920 indicate that mountain goats were introduced to the Olympics cannot be taken as evidence for their absence prior to that time. In the 1920s we still did not have a complete list of mammals native to the Olympics and did not know with any precision the geographic distribution of mountain goats. Post-1920 biologists could have been relying on one of these statements that goats are not native to the Olympics and thereby repeating a possible factoid. Schultz did not take into account the report of naturalists Fannin and Grinnell (1890c), who indicate that the range of mountain goats included the Olympics. Thus her statement that late-nineteenth-century naturalists did not include the Olympic Mountains in the known range of goats is not true.

Schultz (1993:46; 1994:287) writes:

> Because goats were not mentioned in the majority of the cases [she reviews], it does not logically follow that they were not observed or were not present. To assert this would be to succumb to the 'fallacy of negative proof'—claiming that because the observers provided no evidence of goats, they prove that goats weren't there (Fischer 1970). Nevertheless, the considerable amount of negative evidence (failures to mention goats among mammals observed in the Olympics) does not lack significance. By examining the thoroughness and quality of, and the reasons for, the observations of those who *do not* mention goats, a rea-

> sonably accurate assessment can be made of whether or not such observers would have mentioned goats if they had noticed them, and how likely they would have been to have seen them if, in fact, goats had been present. Thus, in comparing and assessing the relative credibility of the various reports, it is necessary to consider carefully (1) the training and experience of the observers, (2) their familiarity with the region and purpose for exploring it, (3) the extent and thoroughness of their exploration activities and opportunities, (4) their familiarity with the habits of wildlife, particularly mountain goats, and (5) the informational detail provided in their observations. When such factors are all taken into consideration, some of the observations clearly assume more reliability than others. (emphasis in Schultz 1993)

Schultz (1993:50–51; 1994:289) states that the personal-journal accounts of both Charles Barnes and James Christie "were printed in full [but] nowhere in these more detailed accounts do either Barnes or Christie mention the sighting of a goat." Schultz (1993:22, 23; 1994: 289–90) variously refers to the "journal accounts of Christie and Charles Barnes" and to their "diar[ies]." However, she also notes that these two men "spent some time in Seattle preparing their narratives for publication, and their accounts were most likely edited by Edmond S. Meany of the [Seattle] *Press*" (Schultz 1993:22; 1994:289). Thus, Schultz's statement that the "personal journal accounts were printed in full" in 1890 is, by her own admission, false.

Robert Wood (1967:vii) writes: "How closely the story as printed in the *Press* [newspaper] conformed to their field notes is a matter of speculation, as the notes were probably destroyed years ago." In his thorough reprinting of the version of Barnes's and Christie's journals originally published by the Seattle *Press*, Harry Majors (e.g., 1981b:185, 187) suggests that (1) the "rough" grammar of some sentences indicates that some portions of the journals were "taken verbatim from [the] original field journal"; (2) "the use of the present tense [in some sentences] indicates a verbatim extract from [the] original field journal"; (3) grammatical and stylistic variations in sentences indicate "obvious editorial insertions"; and (4) some sentences imply "that not all portions of [the] original field journals were actually published in the newspaper narrative" published in 1896. The possibility that mention of a goat sighting was omitted in the published version cannot be acknowledged by Schultz because she claims that the complete journals were published.

Despite Schultz's claims to have assessed the "relative credibility of the various reports," she did not do so in any consistent, thorough, or evenhanded manner. She discredits reported sightings of goats because (1) none of the individuals who reported goats were trained as naturalists or (2) those individuals made natural-history errors. These same two criteria, if applied to the authors of Schultz's "positive statements" that goats were not present in the Olympics, discredit the observations of Webster, Fisher (who reported he saw no pheasants), Grant, and Merriam, who also made natural-history errors.

ETHNOGRAPHIC EVIDENCE

Moorhead and Stevens (1982) did not review the available ethnographic data on the lifeways of Native Americans. That task fell to archaeologist Randall F. Schalk (1993), whose review was not published (see below). He presents a relatively evenhanded and objective review of the ethnographic literature and concludes, correctly in my view, that there is no ethnographic evidence for mountain goats in the Olympic Mountains. There are, however, some points I wish to make about his review. First, Schalk (1993:11, 16) suggests that some of the ethnographic data can be extended back to the "early nineteenth century." Those data, however, come from individual Native Americans who would have been about five years old in 1836; other informants would have been less than ten years old in the middle or late nineteenth century. Because these individuals were speaking from memory to anthropologists about conditions on the peninsula fifty or more years earlier, I have to wonder about the accuracy of their statements.

The potential problem with memory culture, in this case, is exacerbated by the major cultural upheavals that took place late in the eighteenth century and early in the nineteenth century when Native American populations were decimated by introduced European diseases (Dunnell 1991; Thornton 1997). The best estimates indicate that about 80% of the Native Americans present on the Northwest Coast died between 1770 and 1870; over 30% of them died between 1770 and 1800 alone (Boyd 1990). This catastrophic loss of population perhaps meant that few people needed to exploit alpine habitats in order to survive and may have resulted in the loss of anyone who remembered an earlier period of hunting mountain goats in the Olympics. This is why I do not consider the ethnographic evidence to be a solid reflection of late-eighteenth-century cultures. Schalk (1993:42) argues that the ethnographic data he reviewed "indicate local extinction [of mountain goats] must have occurred before the late eighteenth century." Thus, he suggests that the last native mountain goat—if such ever existed—died forty or more years prior to the birth of the few Native Americans who spoke with anthropologists.

Schalk (1993:14–15) quotes William Elmendorf's (1960:197) statement that Indians living on the southeastern portion of the peninsula "bought or traded for most of their goat-wool blankets; the mountain goat did not occur locally. Goat wool was sometimes obtained in trade." He cites Elmendorf's (1960:23) report of a local Native American myth indicating that a particular mythological figure "put no goats in our [Olympic] mountains." There is other affirmative evidence (Elmendorf 1993:96–98) that peninsula people obtained mountain-goat meat and wool from the Cascade Mountains, but Schalk did not cite it because it was published after he completed his report. These data indicate that goats were not present in the Olympics during the nineteenth century and perhaps even in the late eighteenth century (granting some time depth to memory culture), but they do not demonstrate that goats were not present at some more remote time in the past—a point Schalk (1993) recognizes.

Schalk (1993:17) also reviews ethnographic reports of a dog that was kept for its fur, which was woven into garments by native peoples. The "wool dog" was smaller than another breed of dog used for hunting by native peoples. He discusses artifacts related to weaving and suggests that because some of those artifacts appear to be 1,500 years old the wool dog may have been present on the peninsula at that time (Schalk 1993:17, 31–32). These tools may, of course, have been used to weave mountain-goat wool. Schalk (1993:30–31) reviews the prehistoric distribution of the wool dog, but his efforts are thwarted by the fact that this dog has only been positively identified at one site; other archaeological reports of wool dogs that he refers to involve speculative identifications.

Remains of the wool dog dating to the last 800 years have been recovered from the Ozette site on the extreme northwestern corner of the Olympic Peninsula (Gleeson 1970). Paul Gleeson (1970:36) notes that his sample indicates that the remains of dogs are not nearly as abundant during the prehistoric period as during the historic period. Of the fourteen specimens he reports that can both be assigned to the large breed or the small wool dog and also be assigned to the prehistoric or historical period, only two, both dating to the prehistoric period, represent wool dogs. Nine bones and teeth of large dogs date to the historical period and three to the prehistoric period. Gleeson (1970:83) concludes that "the small [wool] dog does not appear in the historic period sample [after 400 years ago]. The need for maintaining dogs for wool was diminishing in the early 1800s because of the introduction of European blankets in the area."

Analysis of the total set of mammalian faunal remains recovered from the Ozette site did not distinguish between the two breeds of dog (Huelsbeck 1983, 1994). However, the abundance of dog remains, irrespective of breed, is significantly lower prior to the historical period than during it. Dog remains make up 1.2% of the total mammal remains recovered from deposits dating between about 800 and 500 years ago, 1.4% of the total mammal remains recovered from deposits dating to about 400 years ago, and 17.6% of the total mammal remains recovered from deposits younger than 400 years. The marked increase in the abundance of dog remains after 400 years ago—essentially the historic period—is inexplicable. Gleeson's (1970) research suggests that this does not represent an increase in the number of wool dogs at the site. If Schalk is correct in suggesting that wool dogs were a surrogate for the mountain goat, then one might expect the abundance of wool dog remains to be higher in the prehistoric than in the historic materials. Unfortunately, data requisite to testing this hypothesis are unavailable. The prehistoric time depth, abundance, and geographic distribution of wool dogs are only now undergoing intensive study (Crockford 1994).

Schalk (1993:7) states that "ethnographic data on the distribution of wool dogs suggest a mutually exclusive relationship to the distribution of mountain goats....[T]he presence of wool dogs on the Olympic Peninsula historically might be interpreted as indirect evidence for the absence of mountain goats in the region....[Thus] the archaeological distribution of wool dogs ... may provide an indirect means of mapping the former distribution of mountain goats." But he does not summarize relevant data to demonstrate the mutually exclusive distributions, and ethnographic data summarized by Rick Schulting (1994) seem to contradict this view. One reference that Schalk cites is Erna Gunther (1927:225), who, in her study of Klallam peoples who lived on the northeast third of the Olympic Peninsula, stated that Native Americans who kept wool dogs lacked access to mountain goats. Three years later, however, Hermann Haeberlin and Gunther (1930:30) stated that various Native American groups on the eastern side of Puget Sound, some of whom hunted mountain goats in the Cascade Mountains, also raised wool dogs. Similarly, groups to the north of Puget Sound in southwestern British Columbia (Suttles 1990a) and groups at the south end of Puget Sound (Suttles and Lane 1990) had wool dogs but also had access to mountain goats. While it may be true that native people on the western side of the Olympic Peninsula had wool dogs because they lacked access to mountain goats, ethnographic data suggest that the distributions of wool dogs and mountain goats overlapped considerably and that the distribution of wool dogs cannot be used to map the distribution of mountain goats.

Schalk's (1993) review of the ethnographic evidence is perhaps the single most convincing bit of information available. That evidence indicates that mountain goats were not present in the Olympics between at least 1850 and 1925 and probably were not present between 1800 and 1850. However, the ethnographic evidence does not tell us if goats were present or absent prior to 1800, as Schalk (1993) agrees. Therefore, given NPS mandates and definitions of native and exotic species, this is insufficient evidence to implement plans to eradicate the goats.

ARCHAEOLOGICAL EVIDENCE

Negative evidence is pernicious in scientific endeavors involving archaeological remains. There are multiple reasons why we may not find archaeological evidence of some prehistoric object or process. It may not have been present in the geographic area where we undertake archaeological research. If it was present, evidence of it may not have been preserved. We may excavate a part of a site that does not contain it. Or we may simply fail to find evidence of it. In all but the first of these situations, the thing or process is not truly absent but only absent from our perceptions. Zooarchaeological research (the study of animal remains recovered from archaeological sites) is plagued by these problems and by others as well.

In the mid-1980s Schalk (1988:i) was hired by Olympic National Park to produce "an archaeological research design and a plan for segmenting Olympic National Park into [archaeological] research/management units" and to perform some archaeological reconnaissance in the park. Schalk (1988:156–57) correctly cautions that before we interpret the zooarchaeological record as indicating changes in the ranges of organisms we must be sure that remains have not been traded by humans or otherwise transported into an area. This is particularly relevant because Schalk (1988:156) also reports that mountain-goat remains were found in an archaeological site on the peninsula early in the twentieth century. Excavating at a site near modern La Push on the Pacific Coast of Washington, Reagan (1917:16) reported finding the remains of "elk, bighorn [sheep], mountain goat, black bear," and other species. But because Reagan added a footnote indicating that the mountain-goat and bighorn-sheep remains were "usually only in the ladle form of horns," Schalk (1988:156), correctly in my view, states:

> In view of the trade in horn spoons reported ethnographically, these occurrences can hardly be interpreted as evidence of the existence of either bighorn sheep or mountain goats *in the Olympics*. In fact, the absence of habitat suitable to bighorn sheep in the Olympics points to trade as the source of the remains of this species. The most likely explanation for the presence of these particular kinds of mountain goat and bighorn sheep remains in [archaeological sites] anywhere on the Olympic Peninsula is that they resulted from trade with other tribes. (emphasis in original)

In a few sentences Schalk has rather eloquently summarized the available archaeological evidence of mountain goats on the Olympic Peninsula that was previously overlooked by Olympic National Park biologists (and myself) and has simultaneously discounted that evidence. Park personnel (e.g., Moorhead and Stevens 1982) may have overlooked some evidence, but that oversight is, in Schalk's view, irrelevant.

Schalk (1988:157) notes that "a critical test for determining the presence or absence of mountain goats during the Holocene in the Olympic Mountains would be to recover a sizable faunal collection from a cave or rockshelter high in the Subalpine Zone. A dry deposit with good faunal preservation and potential for dating the recovered fauna would be ideal. Such a deposit need not contain cultural remains to be useful relative to this question." While we will probably have to excavate more than one site to provide a conclusive answer to the question (Lyman 1995b), I cannot otherwise disagree with Schalk's statement. But if mountain-goat remains are found in that subalpine cave how do we know if they represent native goats or trade items? While far beyond the scope of this chapter, these questions are crucial to the Olympic mountain-goat controversy (see Lyman 1995a).

Zooarchaeological Evidence for Exotic Mammals

Schalk argues that suitable habitat for bighorn sheep is absent from the Olympics. That may be true today, but study of fossil pollen preserved in sediments on the peninsula indicates that the coniferous forests of the Olympic Peninsula were not always so dense and closed as they are today (Gallison 1991, 1994; Petersen et al. 1983). Simply thinning the forest might have made peninsula habitats somewhat more suitable to bighorn, especially if the amount of precipitation was also reduced, which it apparently was during some of the last 10,000 years (see chapter 7). The modern dense forest did not preclude the presence of caribou and another species renowned for being a grassy plains animal—the bison. The bison is not historically reported on the peninsula. Caribou are also unknown on the peninsula in historical times. But Schalk (1988:158) indicates that "bison and caribou are known to have been present on the [Olympic] Peninsula in the late Pleistocene."

The fossil evidence for both species comes from the Manis Mastodon site, located near Sequim, Washington, and first excavated in 1977. The reference Schalk cites in support of his contention that caribou were present prehistorically makes no mention of caribou remains, though that report does note that the remains of "fossil bison" (*Bison* sp.) were found at the site (Gustafson et al. 1979). I am aware of only three sources that state that caribou remains have been found at the Manis site. One says that "bones of...bison and caribou" were found and "bison bones [and] caribou antlers ... were recovered" (Petersen et al. 1983:215, 218). An undated pamphlet probably produced in 1978 or 1979 indicates that the excavators "found a caribou antler" and "bison bones and teeth" and thus "we know that caribou and bison were present" (Gustafson n.d.a). A third statement reads: "A small number of remains representing at least four bison were excavated.... Both cranial and post-cranial bison material has been recovered along with remains of caribou, black bear, and muskrat" (Gilbow 1981:60). Other documents indicate that "bison remains found in the site are from animals that appear to have been butchered [and] several individual bison are indicated" (Gustafson n.d.b; probably produced in 1980 or later) and "remains of at least four individual bison" were recovered (Gustafson 1985:285; written after the 1979 field season). That some caribou and bison remains were found in the first or, less probably, the second (1978) field season is apparent from the following statement: "Caribou and bison were present at Sequim as the last ice age closed" (Kirk and Daugherty 1978:32).

Other archaeologists have found that the caribou antler recovered from the Manis site and the butchered bones of several bison indicate that these species roamed the Olympic Peninsula in late Pleistocene times (Bergland and Marr 1988:17, 24). Carl Gustafson, the excavator of the Manis site, is not included in Schalk's list of archaeologists consulted during the preparation of his report. Does the presence of a single caribou antler (or piece thereof) indicate the native presence of caribou? Even Gustafson seems to have changed his mind regarding the biogeographic significance of the antler. It was apparently found no later than the end of the 1978 field season, but it was not mentioned in the report on the 1979 field season (Gustafson 1985) or in reports apparently produced after the 1979 field season (Gustafson n.d.b). In my view, a single piece of antler, which is often used to make tools by Native Americans in the Pacific Northwest, is hardly affirmative evidence for the native presence of a species. Unfortunately, the final report on the Manis Mastodon site has not yet been completed, so it is difficult to know what the biogeographic significance of the antler might be.

The archaeological remains of bison, caribou, and mountain goats apparently found on the Olympic Peninsula, as described in the literature, indicate only that it is safe to conclude that the bison was native to the peninsula about 10,000 years ago. The bison is represented by teeth, toe bones, and other skeletal parts, and at least four individual bison are represented. It is quite unlikely that those remains were carried to the site from great distances. Only more archaeological research (or perhaps more thorough reporting in the case of the caribou) will clarify the native or nonnative status of the caribou and the mountain goat.

Other Views of the Archaeological Record

Schalk's (1993) report is evenhanded, but it is flawed for two reasons, neither of which is fatal to his conclusion. First, he reviews the archaeological and zooarchaeological evidence for only twenty-three of the thirty-three archaeological sites that had been sampled on the Olympic

Peninsula at that time.[1] Thus, while there is no convincing archaeological evidence for the prehistoric presence of mountain goats even when all thirty-three sites are considered, Schalk's report may be misleading. For example, Paul Crawford, a park biologist, stated after the completion of Schalk's report that "every shred of [archaeological] evidence" had been gathered (quoted in Luoma 1993:41; see also Crawford 1993); clearly, this is untrue. Second, Schalk provides a limited evaluation of the quality of the available zooarchaeological evidence.

In the final paragraph of his report, Schalk (1993:45) offers the following evaluation: "Because of limitations of sample distribution, temporal range, and size, available archaeofaunal [zooarchaeological] data do not allow conclusive statements regarding the presence or absence of mountain goats in the Olympics during the Late Pleistocene or Holocene prior to the 19th century." I could not agree more with this statement (Lyman 1995b). Houston and Schreiner (1994b:10; 1995:206) state that "goat bones have not been identified in remains from archaeological sites on the peninsula." While technically correct—no *bones* have been identified—this statement ignores Reagan's (1917) reported recovery of mountain-goat-horn artifacts from a site on the peninsula. Horn artifacts do not necessarily signify the presence of native mountain goats, but I am not sure why Houston and Schreiner neglect to mention these materials.

The Archaeological Record: Quality and Quantity

As I have indicated elsewhere (Lyman 1994), the zooarchaeological record seems to be the key to resolving the question of the native or exotic status of the mountain goat. But in the late 1980s the requisite kind and extent of archaeological excavation was unclear. Thus I compiled data on how many sites had been excavated, how many mammal bones and teeth had been found, and how much sediment had been excavated at each site in the Olympic Peninsula, the southern Cascades of Washington, and northeastern Oregon. Mountain-goat remains had been found in archaeological sites in the latter two areas, so I reasoned that whatever it took to find goat remains there would probably also be necessary in the Olympics, presuming that the prehistoric abundances of mountain goats in all three areas and the intensity of prehistoric hunting were about the same. In Washington and Oregon mountain-goat remains had been found in arch-

[1] Schalk (1993) actually reviews the evidence from twenty-four sites, but I consider one located on the western edge of Grays Harbor to be too far away from the Olympic Mountains to warrant inclusion.

aeological sites within twenty miles of modern mountain-goat range but not in sites farther away from that range. The other attributes of the three areal samples were more than enough to convince me of the inadequacy of the Olympic Peninsula sample for concluding that mountain goats were never present there prior to 1925.

In the following I consider only archaeological sites located within modern mountain-goat range or located within twenty miles of that range. Forty-eight sites in the southern Cascades and thirty-eight sites in northeastern Oregon have been sampled. In contrast, only seventeen sites have been sampled on the Olympic Peninsula. Over 650 cubic meters of sediments have been excavated in the southern Cascades, and over 900 cubic meters have been excavated in northeastern Oregon. In contrast, only about 85 cubic meters have been excavated on the peninsula. Of the 2,363 bones and teeth of mammals identified from sites in the southern Cascades, only 14 specimens (about 0.5%) have been identified or tentatively identified as mountain goat. Of the 2,471 bones and teeth of mammals identified from sites in northeastern Oregon, only 8 specimens (less than 0.5%) have been identified or tentatively identified as mountain goat. In stark contrast, only 185 bones and teeth of mammals have been identified from sites on the Olympic Peninsula. Thus, the peninsula sample is only about 8% of the size of the other two samples in terms of the number of remains found.

Given the rarity of mountain-goat remains in the southern Cascades and northeastern Oregon samples, I am not surprised that no prehistoric remains of mountain goats have been reported from archaeological sites within twenty miles of modern mountain-goat range on the Olympic Peninsula (see Lyman 1995b for more extended treatment of this issue). But there is a bit more to the story than this. An archaeological site in the north-central Olympics produced a radiocarbon date of about 5,000 years ago and "charred bones" (McNulty 1996:184; see also Schalk 1993:30). I do not know how intensive the efforts were to determine the species of animal(s) represented by those bones because I have not examined the unpublished report on this material said to be on file in the park office. Further, park archaeologist Paul Gleeson told me in late 1993 that the chances of finding bones preserved in the Olympics were slim. Yet he told me during the same conversation about a Native American "burden basket" that had been found high in the mountains (see McNulty 1996:201 for the only published statement on this basket). How, I wondered, could a basket survive the rigors of time but not a bone?

Schalk (1993) also addresses the question of the appropriateness of the available archaeological evidence for assessing the prehistoric geographic range of mountain goats. He reviews archaeological data from sites excavated in northwestern North America in areas where "their prehistoric presence is not debatable" (Schalk 1993:33). How Schalk knows where mountain goats occurred prehistorically is not clear. What he does, in fact, is examine sites within or near the historically documented range of mountain goats. Schalk notes that "while there are undoubtedly some occurrences that have been missed, the sites [discussed] include all that were identified by numerous archaeologists who have worked extensively and intensively in western Washington, coastal British Columbia, and southeast Alaska" (Schalk 1993:33). Based on his interviews and review of written reports, Schalk (1993:36) concludes (1) that sites containing mountain-goat bones tend to be located in areas very near mountains, even if the site is on a saltwater coast, and (2) that "it does appear that the archaeological occurrence of mountain goats decreases rapidly with increasing distance from mountain habitats." The latter is precisely what I had concluded five years earlier (Lyman 1988a).

While I agree with Schalk's conclusion, this portion of his study illustrates some of the problems in attempting to compile all relevant data. For example, Schalk (1993:36) suggests that none of the "sizable [archaeological] excavations on the Fraser River Delta yielded mountain goat remains" and cites three archaeological reports. However, a report he did not cite (Galdikas-Brindamour 1972) indicates that 41 bones and teeth of mountain goat (out of a sample of 1,632 mammal bones and teeth) were found in an archaeological site on the Fraser River Delta.

For the area lying between the approximate latitudes of Seattle, Washington, and Yakutat Bay in south-central Alaska, Schalk (1993) lists nine archaeological occurrences of mountain goats (see table 6.1). Only two of those sites have been reported in published form (Conover 1978; Freed and Lane 1964). One of the seven remaining sites is discussed in an unpublished paper presented at a meeting of professional archaeologists (Savage 1973); faunal data from the other six sites had not been published at the time Schalk wrote, though he was apparently informed about them by the archaeologists he talked to. His list does not include five sites in the area that contain remains of mountain goats.[2]

The faunal data for four of these sites have been published (Cannon

---

[2] After this chapter was written, another site containing mountain goat remains was reported. Stewart and Stewart (1996) identified 624 mammalian remains from the Grassy Bay site, near Boardwalk (see table 6.1), at least to genus. Ten of these represent mountain goat. Stewart and Stewart also studied all faunal remains from Boardwalk, identifying 57 mountain goat specimens among 6,390 total mammalian remains identified at least to genus.

TABLE 6.1.
Archaeological Occurrences of Mountain-Goat Remains in Northwestern North America

Archaeological site	Mountain goat NISP[a]	Total mammal NISP[a]	Comments
Listed by Schalk 1993			
Old Town, Knight Island, Alaska	32	~960	published in Freed and Lane 1964
Boardwalk, Prince Rupert Harbor, British Columbia	4	849	unpublished paper presented at archaeological professional meeting by Savage 1973
FbSx6, Roscoe Inlet, British Columbia	2	7	published in Conover 1978
Ef Sq21, Hopetown, British Columbia	2	?	not published; informant told Schalk
45WH477, Newhalem Rockshelter, Washington	2[b]	?	not published; informant told Schalk (see Mierendorf and Harry 1993)
45WH228, Washington (Ross Lake)	1	?	not published; informant told Schalk
45WH229, Washington (Ross Lake)	1	?	not published; informant told Schalk
45WH303, Washington	21	?	not published; informant told Schalk
Not listed by Schalk 1993			
ElSx1, Namu, British Columbia	4	2528	published in Cannon 1991
Lillooet, British Columbia	3	1843	unpublished master's thesis by Langemann 1987
Kwatna, British Columbia	123	2231	published in Galdikas-Brindamour 1972
DgRr6, Glenrose Cannery, British Columbia	41	1632	published in Galdikas-Brindamour 1972
45SJ1, Cattle Point	1	577	published in King 1950

[a]NISP is the number of identified specimens of bones and teeth.
[b]Some of these bones have been dated to about 1300 years ago (Mierendorf and Harry 1993:43).

1991; Galdikas-Brindamour 1972; King 1950); the faunal data for the fifth is found in an unpublished master's thesis (Langemann 1987).

These oversights on Schalk's part do not invalidate his conclusions (I am sure that I have also overlooked some archaeological occurrences of mountain-goat remains), but they indicate the difficulty of compiling all relevant data. The archaeologists Schalk interviewed clearly were not as familiar with the data as he thought they were. These data typically are scattered among innumerable journal articles and books, and no single library has them all. Some may be unpublished. Statements that "every scrap of evidence was compiled" are seldom true.

At this point two questions must be considered. First, what kind of zooarchaeological (or paleontological) evidence will it take to argue convincingly that mountain goats were present in the Olympics prior to 1925? Obviously, skeletal remains of goats will have to be found. Apparently we need to find prehistoric remains; this seems to be the focus of nearly everyone involved in the controversy (e.g., Crawford in Luoma 1993; Lyman 1995b; Moorhead 1988b; Schalk 1988, 1993). Discovery of prehistoric mountain-goat hair (hair sometimes is preserved in sedimentary contexts where bone is not) probably would not be convincing because mountain-goat wool was traded into the Olympic Peninsula from the Cascade Mountains of Washington by prehistoric and early historical-period peoples (Gibbs 1877; Suttles 1990b). Blood residues on stone tools (e.g., Gurfinkel and Franklin 1988; Hyland et al. 1990; Kooyman and Newman 1992; Loy 1983), which might be identifiable to species, would not be convincing because it could be argued that the tools were used to process hides obtained via trade from elsewhere. If more than one instance of hair, blood residue, or both became available, one might begin to question the conclusion that mountain goats never colonized the Olympic Mountains at some time in the past.

Skeletal remains might come from a paleontological context, without (archaeological) evidence of humans such as artifacts, but in such a case analysis must convincingly show that humans had no hand in transporting the remains to the site. If goat remains are recovered from an archaeological context, the bones and teeth cannot be artifacts, to ensure that they were not traded from elsewhere. Also, the skeletal parts either must not be of high food value (such high value would suggest the possibility of long-distance transport) or must be so numerous as to preclude the possibility of trade or transport. Whether in a paleontological or archaeological context, complete or nearly complete goat skeletons would provide the most unquestionable evidence that the remains were *not* present as a result of long-distance transport or trade. The age of the remains must also be determined.

A second question must be considered: Can we conclude from the zooarchaeological (or paleontological) records that mountain goats were never present in the Olympics? Unfortunately, we cannot because, as I noted in the introduction to this section, there are numerous reasons why we may fail to find prehistoric remains of mountain goats. Once we have devoted the requisite time and effort to looking, as I have suggested, then if we fail to find such remains we can conclude only that mountain goats *probably* were not present. The argument that the absence of zooarchaeological and paleontological evidence of goats is evidence for their absence is presently nonsense, because there has been essentially no effort to find such evidence.

CONCLUSION

Schalk (1993) concludes that one cannot use the available zooarchaeological evidence to argue one way or the other regarding the native or exotic status of the mountain goats on the Olympic Peninsula prior to the nineteenth century (the date is based on ethnographic, not zooarchaeological data). I agree. But I cannot accept Schultz's (1993, 1994) report as a thorough, unbiased, and objective review of the historic evidence.

There seems to be no solid evidence for the presence of mountain goats in the Olympics prior to 1925. The reports of Barnes (1890), Gilman (1896), and Fannin and Grinnell (1890c) are tantalizing but inconclusive. Even so, they cannot be so easily dismissed as Schultz (1993, 1994) or Houston (1995) would have us believe. Schalk's (1993) conclusion is appropriately cautious and underscores the limitations of the archaeological record and the ethnographic evidence. Of course, this is not the way these two reports are presented to the public by park personnel (e.g., Crawford 1993; Houston and Schreiner 1994b, 1995).

Why was Schultz's (1993) report published (Schultz 1994) but not Schalk's (1993)? There seem to be two possible reasons. First, Schalk (1993) concludes that the ethnographic and archaeological data indicate that goats were absent between 1800 and 1925, not resolving the question of whether goats were present sometime between 1492 and 1800. Given NPS mandates to recreate immediately post-Columbian ecosystems (Houston and Schreiner 1994b, 1995; Wagner et al. 1995), if goats were present then the modern goats should be left alone. Second, Park Superintendent Maureen Finnerty once proclaimed that prehistoric data are irrelevant to the controversy. As I note in chapter 7, this calls into question the use of such data by park biologists in their discussions of historical biogeography.

Nine independent scientists were hired by Olympic National Park during the summer of 1993 to review Lyman (1988a), Houston and Schreiner (1993), Schalk (1993), and Schultz (1993). They concluded that there was little more than tantalizing, but clearly very weak and inconclusive, historical evidence that mountain goats were present in the Olympics prior to 1925. These nine reviewers unanimously stated that Schultz's (1993) report was "thorough." Their comments are quoted extensively in the Draft EIS (Olympic National Park 1995). The reviewers, however, were unfamiliar with the primary literature and thus could not form truly objective opinions regarding the historical evidence.

Superintendent Finnerty stated in an April 1, 1994, letter to Keith Gunnar that because the park was "not permitted to photocopy some of the documents [of the primary literature] due to their fragility," the park chose to provide "summaries with detailed reference lists [as] a professionally acceptable means for conducting such reviews." The reviewers did not have the time to read the primary literature: they were given three weeks to read and prepare their reviews of four documents totaling about 170 pages. It is not surprising that the reviewers did not discover the flaws in these reports that I have outlined above. Nor is it surprising that they discarded the hypothesis that mountain goats might have been present in the Olympics at some time prior to 1925 in favor of the park's hypothesis that goats were never present.

The park-sponsored reviews of the historical, ethnographic, and archaeological data have convinced various professionals not directly affiliated with the NPS, Olympic National Park, or the controversy that the extant mountain goats are exotic (e.g., Hutchins 1995). This may well be the correct conclusion, but, as I have shown here, the evidence mustered in support of that conclusion is flawed and contradictory.

Plate 1. An Olympic Peninsula mountain goat, with Mt. Angeles in the background. Photo by Antje Gunnar.

Plate 2. A mature and a young Olympic mountain goat.
Photo by Antje Gunnar.

Plate 3. The Constance Pass area of Olympic National Park, where goats were first reported in 1930 and 1935. Photo by Antje Gunnar.

Plate 4. Mt. Angeles (upper left) and Klahhane Ridge area of Olympic National Park, where goats were first reported in 1933. Photo by Keith Gunnar.

Plate 5. Olympic National Park mountain goats.
Photo by Keith Gunnar.

Plate 6. Dust-bathing in a wallow in Olympic National Park, showing mountain goat as an erosional force. Photo by Keith Gunnar.

Plate 7. The next generation.
Photo by Keith Gunnar.

Plate 8. Truly, a beast the color of winter.
Photo by Keith Gunnar.

7

ON THE HISTORICAL BIOGEOGRAPHY OF THE OLYMPIC PENINSULA

In this chapter I review some of the biological evidence that park biologists have called upon to argue that mountain goats were not present in the Olympic Mountains prior to 1925. Much of this evidence is found in the unpublished manuscript "Biogeography of the Olympic Peninsula" (Houston and Schreiner 1993) by Douglas B. Houston and Edward G. Schreiner, two park biologists. The manuscript was edited and published in the 1994 monograph on Olympic mountain goats (Houston, Schreiner, and Moorhead 1994); the published form (Houston and Schreiner 1994a) lists three authors (Houston, Schreiner, and N. M. Buckingham) even though it is very similar to the unpublished form.

Although Houston and Schreiner (1993, 1994a) summarize much of what is known about the modern distribution of plants and animals on the Olympic Peninsula, they do not really explore the *historical* aspects of the biogeography that are critical to the controversy. Instead, they reiterate some of the same arguments that have been made since Scheffer's (1946) original descriptive summary of Olympic Peninsula mammalogy to show that mountain goats were never present there prior to their 1925 introduction. As is typical of scientific inquiry, the document raises almost as many questions as it answers, some of which could be critical in determining the native or exotic status of mountain goats in the Olympics.

BIOGEOGRAPHIC THEORY

Biologist James H. Brown (1971, 1978) suggests that the patchy or discontinuous distribution of various mammals that are presently found only on the tops of Nevada's mountain ranges could result from colonization of those mountains and intervening desert lowlands during the ice age or Pleistocene (see also Findley 1969). After the Pleistocene ended about 10,000 years ago, various mammal populations became isolated on the mountains as the surrounding lowlands took on their modern, desertlike form. Donald K. Grayson (1977, 1981, 1982, 1987, 1993; Grayson and Livingston 1993), a zooarchaeologist known for his studies of the prehistoric zoogeography of mammals in Nevada and adjacent areas, has reported zooarchaeological and paleontological evidence that confirms Brown's hypothesis.

Brown attempts to show that the theory of island biogeography proposed by Robert H. MacArthur and Edward O. Wilson (1967) might not account for the modern distribution of the Nevada mammals. MacArthur and Wilson suggested that the organisms found on oceanic islands were a result of the interaction of colonization and extinction events. Oceanic islands, as later defined (MacArthur 1972:108), are those that never had a terrestrial connection with the mainland. As William Newmark (1987:430) states: "Land-bridge islands are islands that were formerly connected to the mainland and were created by a rise in the level of the ocean." Faunas found on both sorts of island are usually referred to as insular. MacArthur and Wilson were largely concerned with animals on oceanic islands. They reasoned that extirpations or local extinctions on oceanic (and landbridge) islands would be compensated for by new colonizations in direct relation to the island's distance from the mainland and its size. The farther away and the smaller an island was, the less likely it was that new colonization events (from the mainland) would offset new extinction events on that island.

A more modern, revised form of island biogeography (what I here term "insular biogeography") concerns landbridge islands. Bruce Patterson (1987:324) writes that "landbridge islands were directly connected to species-rich mainland areas during glacial episodes of the Pleistocene, and they presumably shared their diverse biotas. Since their disjunction from the mainland areas, landbridge islands are thought to have suffered a net loss of species via local extinction [or what is called] (faunal relaxation)." Landbridge islands should, therefore, at least initially, contain all or most of the species of plants and animals that are found on the mainland. But when the glaciers melt and sea level rises, the islands become isolated from the mainland and thus the animals and plants on them become isolated from their larger mainland populations. As the islands grow smaller with rising sea levels, species are extirpated or die out as habitable area and habitat diversity decrease. Of course, when speaking of Nevada, the "islands" are isolated habitats on mountaintops and slopes whereas the "oceans" are lowland habitats in basins between the mountains.

Thus Brown proposes, and Grayson shows, that a modified form of the theory of island biogeography, insular biogeography, accounts for the mammals on Nevada mountaintops. For example, the pika (*Ochotona princeps*), a small rabbitlike creature that lives only in alpine environments, today is found on only five of the nineteen mountain ranges in Nevada and surrounding areas and not in lowland areas between those ranges. Zooarchaeological research has shown that the pika at some time in the past occupied at least four additional mountain ranges and also some of the lowland areas where it cannot survive today (Grayson 1993; Mead 1987). Because the Nevada mountain ranges contained relatively isolated mountaintop habitat islands since the end of the Pleistocene, pikas were unable to recolonize those mountain ranges where they became locally extinct over the last 10,000 years. The general implication of such findings is clear: the present distributions of mammals are not always good keys to the prehistoric distributions of mammals. In the last decade a growing body of evidence has shown that insular-biogeographic theory works well as an explanation for the distributions of many mammals and birds in the western United States and elsewhere (e.g., Brown 1986; Davis et al. 1988; Lomolino 1993, 1994; Lomolino et al. 1989; Newmark 1986a, 1986b, 1987; Patterson 1984, 1987, 1990; Patterson and Atmar 1986).

The Nevada mountaintop mammal research suggests this question: Did mountain goats colonize the Olympic Mountains during the Pleistocene, only to become extinct sometime after the end of the ice age? If so, according to the modern theory of insular biogeography, then mountain goats probably could not have recolonized the Olympics during the last 10,000 years, the geological epoch known as the Holocene. Given these observations, perhaps the perception of park biologists that mountain goats were *never* naturally present in the Olympics is incorrect.

Moorhead and Stevens (1982) conclude that the mountain goats of the Olympics were "nonnative" without detailed consideration of the implications of modern insular-biogeographic theory. They indicate that as glaciers withdrew at the end of the Pleistocene and the Holocene began, "the climate warmed [and] goat distribution has been progressively restricted to high mountain retreats in the Pacific Northwest" since then (Moorhead and Stevens 1982:47). This fits well with the theory of insular biogeography. Importantly, this statement is virtually the only aspect of the biogeographical theory that anyone involved with the controversy mentioned at the time, with one irrelevant exception.

Hutchins and Stevens (1981:59), for example, merely note that "the absence of native goat populations in the Olympic Mountains poses an interesting biogeographical question, particularly since the habitat is ideally suited to these animals and since they are native to the Cascade Range less than 100 miles to the east." They also argue that the Olympic Mountains had long been isolated (in effect, a biogeographic island) and suggest that this isolation took place well before mountain goats colonized the North American continent (Hutchins and Stevens 1981:61). Were such isolation to have been constant, mountain goats would not have been present in the Cascade Mountains or anywhere else in North America prior to the isolation event. Thus this statement is irrelevant to the biogeographic history of mountain goats, yet it is regularly repeated.

Biologist Victor B. Scheffer's (1993a, 1993b) perspective on what he terms "The Olympic Goat Controversy," while in ways unique, emanates from the rhetoric of Olympic National Park biologists. For example, he uses the park biologists' map of the "native range" of mountain goats as his first figure; the Olympic Peninsula is not included in the native range, but neither are areas to which goats clearly are native (Lyman 1994); Scheffer and the park biologists do not tell us how the map was derived. Scheffer also echoes a statement found in Olympic National Park's (1987a) mountain-goat-management planning document. Scheffer (1993a:918) writes: "Geologic clues indicate that continental ice in the Puget Sound Basin would have isolated the high Olympics from the high Cascades long before the first goats reached North America, perhaps 40,000 years ago." Compare this with the statement found in Olympic National Park (1987a:17): "geologic and glacial events ... isolated the Olympic Mountains from other ranges long before the ancestral goats entered North America across the Bering land bridge [perhaps] 40,000 years [ago]." These almost identical statements are, as I argued above, irrelevant to the question of whether mountain goats ever naturally colonized the Olympic Mountains.

Largely in response to Moorhead and Stevens's (1982) research, I published a paper (Lyman 1988a) in which I argued that the lack of evidence for pre-1925 goats was not necessarily evidence for their absence. Virtually all twentieth-century scientists who had written on the mammals of the Olympics had flatly denied the possibility that mountain goats could have been in the Olympics at some time in the past. Early-twentieth-century knowledge had become dogma in the late twentieth century. Because of this belief, no new knowledge was being generated. I wrote that "the conclusion that mountain goats are not native to the Olympic Peninsula is based on negative evidence" (Lyman 1988a:15).

GETTING MOUNTAIN GOATS INTO THE OLYMPICS

There are two important aspects to using insular biogeographic theory to build a model to show how mountain goats might have occupied the Olympic Mountains naturally at some time in the past. One concerns the history of Pleistocene glaciation of the Pacific Northwest; the other concerns vegetation.

The dates typically assigned to the Wisconsin glacial period (the last of several that occurred during the Pleistocene) extend from approximately 70,000 years ago until about 10,000 years ago (e.g., Easterbrook 1986). As we have learned more about the history of the Cordilleran ice sheet (one of two continental ice sheets that covered much of the northern portions of North America during the Pleistocene), it has become increasingly difficult to maintain a single name for a dynamic geological period. For example, there were at least three major periods of glacial advance into Puget Sound over about the last 200,000 years (Easterbrook 1992). I have retained the terminology "Wisconsin glaciation" for simplicity's sake. The "late Wisconsin" glaciation referred to by Houston and Schreiner, also known as the Fraser glaciation, dates to between about 22,000 and 10,000 years ago; including the interglacial period prior to the late Wisconsin glaciation extends its beginning date back to about 30,000 years ago (Easterbrook 1986). The Cordilleran ice sheet extended from southern Alaska in the north to northern Washington in the south and from the Pacific coast on the west to the eastern foothills of the Rocky Mountains on the east.

Northern areas of Washington, Idaho, and western Montana were covered with the Cordilleran ice sheet during the Wisconsin glacial period 15,000 ± 1000 years ago; the Cascade Mountains of western Washington were covered with alpine glaciers at this time (see fig. 7.1). The glaciers were retreating from northeastern Washington and northern Idaho by 13,000 years ago. The Puget lobe of the Cordilleran ice sheet, which occupied present-day Puget Sound, reached its southern limit about 14,250 ± 250 years ago and shortly afterward began to recede. By 13,600 years ago its southern terminus had reached the northern end of

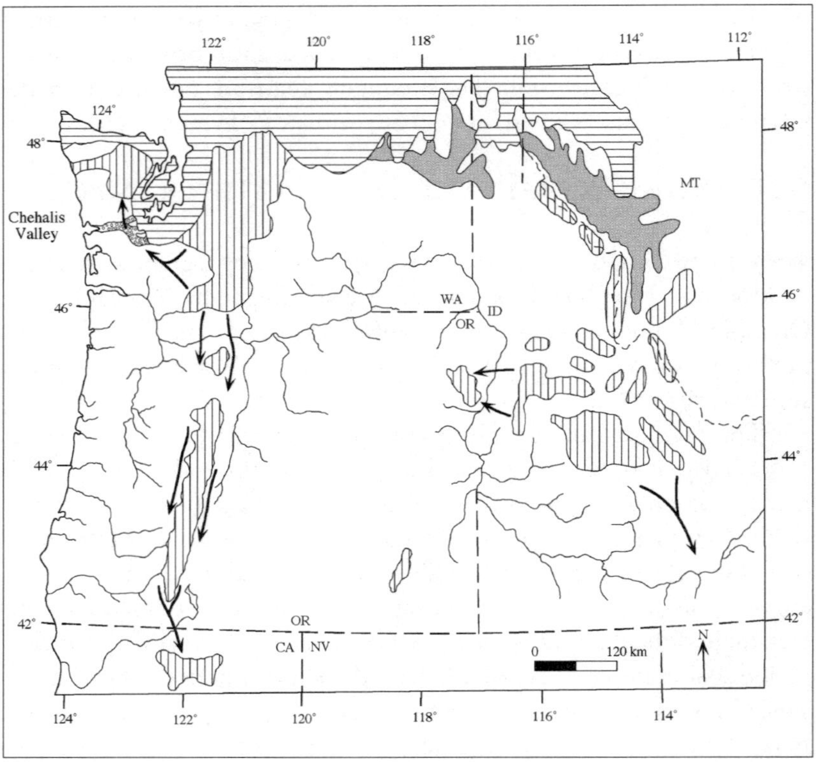

Figure 7.1. Extent of glacier ice about 15,000 years ago (incomplete for western Montana). The Cordilleran ice sheet is indicated by horizontal cross-hatching; alpine glaciers are indicated by vertical cross-hatching; stipling indicates glacial lakes. Alpine glaciers of the Olympic Mountains would not be present at this time, having receded by 18,000 years ago. Note the location of the Chehalis Valley glacial outwash plain. Heavy arrows indicate the hypothesized dispersal routes of mountain goats. See also figure 7.2.

Puget Sound (Booth 1987; Easterbrook 1986, 1992; Waitt and Thorson 1983). Alpine glaciers "had retreated at the time of the culminating advance of the Puget lobe [14,250 years ago]. In many valleys west of the Cascade crest, lakes were ponded between thick ice in the Puget lowland and the Cascade alpine glaciers" (Porter et al. 1983:84). The alpine glaciers of the Olympic Mountains had advanced and retreated prior to 18,000 years ago (Waitt and Thorson 1983).

On one hand, much of the Washington Cascades today inhabited by mountain goats and other alpine mammals was unavailable for occupation between approximately 17,500 and 13,500 years ago (see figs. 7.1 and 7.2). Alpine habitats of the Olympic Mountains, on the other hand,

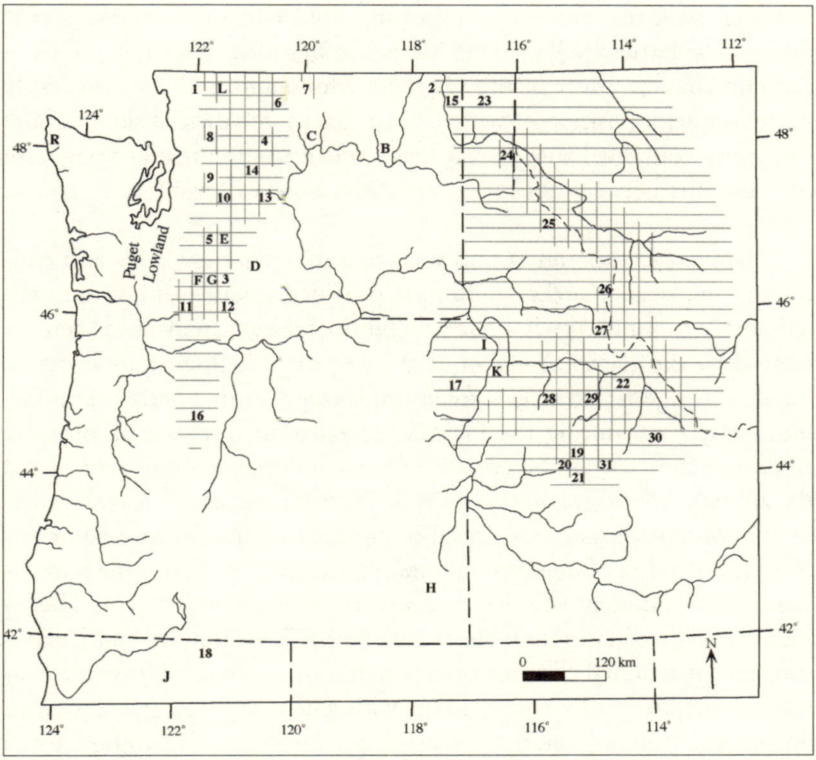

Figure 7.2. Modern distribution of mountain goats in the Pacific Northwestern United States. The horizontal cross-hatching is the distribution shown by Hall (1981); the vertical cross-hatching is the distribution shown by Johnson (1977; see also Wigal and Coggins 1982). Numbers represent historically reported locations of mountain goats (Montana not included) and letters represent reported prehistoric records of mountain goats (see table 7.1 for a key). Compare this figure with figure 7.1.

would have been deglaciated and available for occupation during this period as the Olympics were relatively ice free after 18,000 years ago.

Fossil pollen data indicate that between 17,000 and 12,500 years ago the area between Puget Sound and the Cascade crest contained a vegetational gradient. Treeless conditions prevailed in the north, pine-parkland in the mid-latitude area, and forests in the south. Dominant vegetation in the mid-latitudes consisted of pine (*Pinus* sp.), spruce (*Picea* sp.), mountain hemlock (*Tsuga mertensiana*), and herbs (Heusser 1983). In the Puget lowland tundra-parkland vegetation prevailed between 19,000 and 17,000 years ago and changed to spruce-pine parkland between 17,000 and 15,000 years ago. From 15,000 to 12,500 years ago spruce–mountain hemlock parkland prevailed, suggesting mild climatic conditions (Barnosky 1984; Barnosky et al. 1987; Mathewes 1993; Whitlock 1992). Due to the rain shadow effect of the Olympic Mountains and the apparently more southerly position of winter storm tracks, the eastern slopes of the Olympics contained some open herb-and-shrub-dominated vegetation communities between at least 12,000 and 11,000 years ago (Petersen et al. 1983).

Between 17,500 and 13,000 years ago the Puget lowlands and contiguous unglaciated areas would have provided suitable (if not optimal) habitat for mountain goats. The suitability of the area may have been enhanced by the glacial omission of more optimal high-altitude habitats. C. R. Harington (1971:1091), for example, suggests that periods of maximum glaciation during the Pleistocene eastward dispersal of ancestral *Oreamnos* from eastern Siberia into Alaska and thence southward forced the animals "to survive on the lower grasslands" and that "an analogy between winter-summer and glacial-interglacial mountain goat habitat is relevant." Studies of modern mountain goats indicate that some populations utilize different summer and winter ranges, summer ranges being found at higher elevations when winter ranges have minimal south exposure and/or minimal cliff area from which snow is removed by wind. Seasonal variation in diet concomitant with shifts in ranges has also been documented among modern goats (e.g., Hebert and Turnbull 1977; Smith 1977). During the Wisconsin glaciation, then, the relatively snow- and ice-free eastern slopes of the Olympics might have been quite attractive to immigrating mountain goats.

Dalquest (1948) speculates that prior to the last major glacial advance (the Wisconsin) about 18,000 years ago mountain goats occupied an extensive range in Washington. When the last glacial advance occurred (17,500 to 13,500 years ago), mountain goats were confined to the southern Cascades of Washington. Dalquest (1948:109) thus writes that goats "seem not to have crossed the Columbia River to become established in the Cascades of Oregon," but fossils of goats from Mount Shasta in northern California suggest this may be incorrect (see fig. 7.2). Dalquest believes that as glacial ice retreated mountain goats dispersed northward through the Cascades of Washington and back into British Columbia. This is a reasonable hypothesis.

The major obstacle to the dispersal of mammals from the southern Cascades of Washington to the Olympics may well have been the glacial outwash plain of the Chehalis River (see fig. 7.1). As the Puget Lobe of the Cordilleran ice sheet filled Puget Sound, water pooled south of the lobe could not drain to the north as it does today. Instead, this water spilled southward and westward into the Chehalis Valley and then into the Pacific Ocean (Booth 1987:85–86). This valley may have been crossed at some unknown time in the past when filled with glacial-melt water by the yellow-pine chipmunk (*Tamias amoenus*), a species found in both the Olympics and the Cascades today but not in intermediate areas, indicating that the Chehalis Valley and the Puget lowland may have not been insurmountable barriers to the dispersal of many mammalian taxa. Recall that forty species are common to both sets of mountains.

Figure 7.2 shows two interpretations of the modern distribution of mountain goats, many of the historically reported locations, and archaeological sites containing remains of mountain goats in the northwestern United States. References used to construct this figure and location names are listed in table 7.1. These historic and prehistoric data are readily accounted for by the model of mountain goat dispersal summarized in figure 7.1, although in some cases it is unclear if the prehistoric remains of goats in sites in Washington and Oregon represent specimens that were transported great distances by humans. The model in figure 7.1, with the exception of goat dispersal into the Olympics, is similar to one constructed independently by Jim Mead (1983), a Quaternary vertebrate paleontologist. Note that my model has implications for other mammalian taxa such as the wolverine (*Gulo luscus*), pika, and grizzly bear, which are historically unknown in the Olympics but could have dispersed to the Olympics at some remote time in the past only to become extinct later (I return to these mammals below). Finally, while my model could account for the prehistoric presence of goats in the Olympics, it does not mean that goats were there prior to 1925. Thus I cannot agree with Ernest Lundelius et al. (1983:344), who indicate that "the late Pleistocene faunas of the Olympic Peninsula were similar to those of the Rocky Mountains to the east; the mountainous parts of both regions shared taxa such as *Oreamnos americanus*." There are, as yet, no fossil or zooarchaeological data to substantiate such a claim.

MISSING MAMMALS

Park biologists are cognizant of the apparent biogeographic isolation of the alpine habitats under their care. Scheffer (1946:5), for example, writes: "Many species now living in the Cascades have been unable, in

TABLE 7.1.
Historic and Prehistoric Records of Mountain Goats in the Northwestern United States (see figure 7.2)

State Reference	Location name (H, historic; P, prehistoric)	Figure 7.2 symbol
Washington		
Brooks (1930)	Mount Baker [H]	1
Scheffer and Dalquest (1944)	Deep Lake [H]	2
Booth (1947)	Goat Rocks [H]	3
	Lake Chelan, Canoe Creek [H]	4
	Methow River [H]	4
	Palmer Lake [H]	4
	Mount Rainier [H]	5
	Billy Goat Pass [H]	6
	Mount Chopaka [H]	7
Dalquest (1948)	Mount Baker [H]	1
	Mount Higgins [H]	8
	Mount Index [H]	9
	Deception Creek (Goat Basin) [H]	10
	Mount Saint Helens [H]	11
	Mount Adams [H]	12
	Mount Stuart [H]	13
	Lake Chelan [H]	4
	Stevens Pass [H]	14
Johnson (1961)	Sullivan Lake [H]	15
Campbell and Johnson (1983)	Chopaka Mountain [H]	7
Reagan (1917)	LaPush [P]	A
Collier et al. (1942)	Site 24 [P]	B
Grabert (1974)	Coulee Creek Rockshelter [P]	C
Gustafson (1983)	Wild Rose Rockshelter [P]	D
Gustafson (1983)	Fryingpan Rockshelter [P]	E
Daugherty et al. (1987a)	Judd Peak Rockshelter [P]	F
Daugherty et al. (1987b)	Layser Cave [P]	F
Lyman (1995a)	Deceptive Beauty Rockshelter [P]	G
Schalk (1993)	Newhalem Rockshelter [P]	L
Oregon		
Grant (1905)	Mount Jefferson [H]	16
	Wallowa Mountains [H]	17
Bailey (1936)	Mount Jefferson [H]	16
Hubbard (1967)	Rattlesnake Creek [P]	H
Leonhardy and Thompson (1991)	Camp Creek [P]	I
Northern California		
Johnson (1970)	Lava Beds National Monument [H]	18
Sinclair (1903, 1904, 1905)	Potter Creek Cave [P]	J
Furlong (1906)	Samwell Cave [P]	J

State Reference	Location name (H, historic; P, prehistoric)	Figure 7.2 symbol
Idaho		
Davis (1939)	Loon Creek [H]	19
	Stanley Lake [H]	20
	Alturus Lake [H]	21
	Boulder Peak [H]	21
	Salmon River Mountains [H]	22
Rust (1946)	high peaks of northern Idaho [H]	23
Brandborg (1955)	Selkirk Mountains [H]	23
	Cabinet Mountains [H]	24
	North Fork, Clearwater River [H]	25
	Lochsa River [H]	26
	Selway River [H]	27
	Salmon River [H]	22
	South Fork, Salmon River [H]	28
	Middle Fork, Salmon River [H]	29
	South Fork, Payette River, and Boise River [H]	20
	Big Lost River [H]	30
	Big Wood River [H]	31
Randolph and Dahlstrom (1977)	Bernard Creek Rockshelter [P]	K

Note: Records in western Montana are not included.

the time that has elapsed since the Ice Age, to cross the forested lowlands of the Puget Sound trough and occupy suitable habitats in the Olympics." Scheffer (1993a:916), citing Newmark (1987), later wrote that the Olympic Mountains represent a "land-bridge island" that was goat free during the development of the soils and biotas of the park. This statement was apparently reviewed by Olympic National Park biologists prior to its publication (Scheffer thanks them in the acknowledgments); thus I conclude that those biologists are in agreement with it. But in using the term "land-bridge island" Scheffer is indicating that the Olympic Mountains have not always been biogeographically isolated from the southern Cascades mainland. What park biologists (e.g., Moorhead 1988a; Moorhead and Stevens 1982; Sleeper and Hutchins 1989) and others who cite them (e.g., Carlquist 1990; Luoma 1993) are implying in their statements is *continuous*, rather than intermittent, biogeographic isolation of the Olympics, in which case they should use the term "oceanic island." Oceanic islands "have always lacked dryland connections to potential source pools [of species] and are often relatively impoverished [in terms of number of species], supporting only highly vagile species" (Patterson 1987:324; see also MacArthur 1972).

Newmark (1986b:95) notes that "within 14 of the 24 western North American national parks included in [his] study with the best historical current species sight records (average park age = 75.1 years), there have been only three post-establishment [of the park] natural colonizations by species of lagomorphs, carnivores and artiodactyls not previously found within the parks." A similar pattern of historic losses from and additions to Mount Rainier National Park's mammalian fauna is documented by A. R. Weisbrod (1979; Weisbrod and Dragavon 1979) and Susan Dixon (1996). It seems reasonable to conclude that establishment of Olympic National Park in 1938 and concomitant modern development of intervening lowlands have probably significantly decreased the chances that mountain goats would naturally colonize that park from the southern Cascades during the twentieth century.

The pattern of oceanic islands not having the large number of species found on the adjacent mainland represents what I refer to as the "missing mammals" problem in Olympic National Park. Park biologists reason that goats are nonnative in part because several species of mammals historically found in the Cascade Mountains of Washington have never been historically reported in the Olympic Mountains. As noted in chapter 6, a complete list of mammals occupying the Olympics was unavailable until about 1940. The point here is that basing biogeographic conclusions on missing mammals raises the problem of deciding if a species is truly absent or was merely missed by those doing biological surveys.

In his synopsis of Olympic Peninsula mammals, Victor Scheffer (1946:5) lists eleven species, including the mountain goat, historically found in the Cascade Mountains that he indicates "might logically be expected in the Olympics but which do not occur there."[1] Nearly fifty years later he wrote that "the goat is one of 11 species of mammals native to the Cascade Range of Washington which are *not* native in the Olympic Range only 120 km [75 miles] away" (Scheffer 1993a:918). These mammals are listed in table 7.2. Scheffer's observation was not without precedent. As indicated in chapter 6, E. B. Webster (1920:146) had earlier noted the "peculiar" absence of mountain goats from the Olympics. Still earlier, C. Hart Merriam wrote that "sheep, goats, porcupines, coyotes,

[1] Scheffer's 1946 manuscript (including its title) was revised in 1949 and scheduled for publication in 1950. It was not published, however, until 1995 as number two in the monograph series of the Society for Northwestern Vertebrate Biology (Scheffer 1995). In that published version, the quotation is found on page 15.

foxes, [and] grisly bear" (quoted in Schultz 1993:40; 1994:283) were "said" to be absent from the Olympics.

Why should these mammal species be missing in the twentieth century? In an unpublished manuscript Merriam suggests a possible answer that was the only one considered until 1988:

> The gap that separates the Olympic Mountains from the Cascade Range, the only other mountains of the region, is only about a hundred miles wide and is filled by dense forest . . . affording continuity of range for species inhabiting the mixed Transition and Canadian zones,[2] but the higher parts of the Olympics, falling within the Hudsonian and Alpine zones, have been disconnected from corresponding zones to the north and east for at least several million years, a period long enough to admit a considerable amount of differentiation in the species stranded here. . . . Contrasting the faunas and floras of the two regions, some highly interesting differences appear. The flora of the Olympics is rich and varied and comprises a number of types not known from the Cascade Range. The fauna, on the other hand, at least as far as the land vertebrates are concerned, is rather meager and is not known to include a single *specific* type not found in the Cascades. . . . Among mammals, perhaps the most striking peculiarity of the region is the species it lacks. There are no mountain sheep, goats, porcupines, coyotes, foxes, or grizzly bears in the Olympics. (quoted in Schultz 1993:40–41, 1994:283; emphasis in original)

Merriam suggests that biogeographic isolation of high-altitude habitats of the Olympic Mountains was created by low-lying intervening area covered with dense forests. According to Merriam, this area, known as the Puget Lowlands, prevented some animals from ever entering the Olympics from the Cascade Mountains. It is this *modern* isolation that is called on today by park biologists to account for the missing mammals in table 7.2.

[2] Merriam is here referring to what he called "life zones." For Merriam, the distribution of plant and animal life was dependent on environmental factors such as temperature, precipitation, topography, and vegetation. He modeled biogeographic regions as a series of east-west-trending bands across North America and called these life zones, noting as well that many of these same bands could be found on mountains. From south to north (and from low to high elevation on a mountain), the life zones included the Tropical, Austral, Transition, Canadian, Hudsonian, and Arctic (or Alpine). This system for organizing biogeographic data has since fallen from favor and is seldom used today.

TABLE 7.2.
Missing Mammals

Taxon	Merriam unpublished journal	Scheffer 1946	Johnson and Johnson 1952	Hutchins and Stevens 1981[a]	Moorhead and Stevens 1982	ONP 1987a	Moorhead 1988a	Houston and Schreiner 1993, 1994a
Ochotona princeps (pika)	+	+	+	+	+	+	+	+
Spermophilus saturatus[b] (golden-mantled ground squirrel)	+	+	+	+	+	+	+	+
Microtus richardsoni (water vole)	+	+						+
Synaptomys borealis[c] (northern bog lemming)		+						+
Erethizon dorsatum (porcupine)	+	+	"rare in southern peninsula"					"occasional, no established population"
Canis latrans (coyote)	+	"recent immigrant"						"recent immigrant"
Vulpes vulpes (red fox)	+	+	"introduced"	+		+		+
Ursus arctos (grizzly bear)	+	+	+			+	+	+

Gulo luscus (wolverine)		+				+	+	+
Lynx canadensis (lynx)		+	+			+	+	+
Oreamnos americanus (mountain goat)	+	+	"introduced"	+	+	+	+	+
Ovis canadensis[d] (bighorn sheep)	+	+			+	+	+	+

Note: Species indicated by "+" are listed by author(s) heading the column as species not found in the Olympic Mountains but historically present in the Cascade Mountains of Washington. Sleeper and Hutchins (1989) indicate the list consists of "seven terrestrial mammal species" plus the mountain goat, but do not indicate the species.
[a] Hutchins and Stevens (1981) indicate the list consists of eight species, but list only the four indicated.
[b] Houston and Schreiner (1993) list this species as *S. lateralis,* but the general consensus is that *S. lateralis* and *S. saturatus* are separate species (e.g., Bartels and Thompson 1993; Hall 1981; Tomich 1982; Trombulak 1988).
[c] The northern bog lemming historically occurs only in the northernmost Cascades of Washington (Wilson et al. 1980).
[d] The bighorn sheep is historically known only from the northernmost Cascades of Washington and much of eastern Washington, well east of the eastern foothills of the Cascades (Johnson 1983).

Published and unpublished papers on Olympic National Park continue to include the list of mammals missing from the park as evidence that goats were never present (e.g., Houston and Schreiner 1993, 1994a; Moorhead 1988a; Moorhead and Stevens 1982; Olympic National Park 1987a, 1995:68; Sleeper and Hutchins 1989). Those who argue that mountain goats are not native to the Olympics typically fail to note the forty nonvolant species common to both mountain ranges (using species lists of Johnson and Johnson 1952 for the Olympics and Schamberger 1970 for Mount Rainier in the southern Cascades). This fact weakens the characterization of the Olympic Mountains as biogeographically isolated and inaccessible to mountain goats.

Did all twelve species actually have access to the Olympics? Perhaps not. The northern bog lemming today is rarely found, and only in the extreme northernmost Cascades of Washington (Wilson et al. 1980), so perhaps it should not be included with the other eleven species, which are found today in the southern Cascades (e.g., around Mount Rainier or farther south). Similarly, there is no undisputed historical record of the bighorn sheep being in the southern Cascades; it is historically known to occur only in the northern Cascades (Johnson 1983). Prehistoric specimens of bighorn sheep (Lyman and Livingston 1983) and northern bog lemming (Rensberger et al. 1984; Rensberger and Barnosky 1993) have been found in the southern Cascades. But during the historical period these two species clearly did not have the same direct access to the Olympics as species such as the pika and red fox.

The length or taxonomic richness of any list of species depends largely on the definition of a geographic area and the rules for inclusion. (Are only nonvolant taxa included? Are historical records included and prehistoric records excluded?) Scheffer and the Olympic National Park biologists have never explicitly stated how they compiled the taxonomic lists of missing mammals to which they refer regularly. One exception to this is Houston and Schreiner (1993:5, 1994a:32), who indicate that while the porcupine has "been reported very rarely during the past few decades [in the southern portion of the Olympic Peninsula], apparently a population has not become established [on the peninsula]." Here they are implicitly using the criterion of an established population rather than a few reports of a vagrant or lost individual to define the geographic range of a species (see Gaston 1991 and references therein for a discussion of defining geographic ranges of organisms). The other exception (Houston 1995) came after a direct challenge was made to these data (Lyman 1995a).

Perhaps because no rules for building the lists in table 7.2 have been made explicit, those lists are quite varied. For example, Johnson and Johnson (1952) do not include the northern bog lemming and bighorn sheep in their list of missing mammals, probably because these two species were not historically known to occur in the southern Cascade Mountains of Washington. Houston (1995) indicates that the northern bog lemming should be included (and would probably include the bighorn sheep) because its range was more southern during the Pleistocene. This admits that the modern record tells us little about the past and that the list in table 7.2 will probably be considerably modified as we learn more. To include species on the basis of the fossil record would force us to consider the possibility that many species might have been present in the Olympics at an earlier time, but became locally extinct sometime during the last 10,000 years. This possibility is why I place minimal faith in the list of mammals missing from the Olympics as compelling evidence for the absence of mountain goats from the Olympics prior to 1925.

Houston and Schreiner (1993:3; 1994a:29) include the white-tailed deer (*Odocoileus virginianus*) as a member of the Olympic Peninsula mammalian fauna because, while it is not presently found there, the species did occur there around the turn of the century. It is clear from the archaeological record that this species had a much different biogeographic range in the Pacific Northwest during the nineteenth century, and in the prehistoric past, than it does today (Livingston 1987). Including it on the list while acknowledging its recent local extirpation apparently does not suggest to Houston and Schreiner that the same fate may account for the apparent late-nineteenth-century absence of mountain goats and other species in table 7.2 from the Olympic Peninsula.

By relaxing the criteria for inclusion of species in the list of missing mammals, and therefore including the northern bog lemming and the bighorn sheep, park biologists have effectively lengthened the list of missing mammals by 20%, which makes it seem more convincing that a significant subset of Washington's mammals never colonized the Olympics. The park biologists' list is 100% longer than the only list compiled by a nonpark biologist: park biologists consider twelve species to be missing whereas Johnson and Johnson (1952) list only six. The difference here seems to be a result of varied criteria and rules for defining the set of "missing mammals."

The irrelevance of the lists in table 7.2 for understanding the historic biogeography of the area is underscored by the prehistoric caribou and bison remains that have been found on the Olympic Peninsula (Gustafson 1985; Petersen et al. 1983). Neither of these species is mentioned in any document about the goats written by park personnel. I believe that at least the bison was present on the peninsula at the end of the Pleistocene or about 10,000 years ago. The caribou remains (apparently only antler) may have been traded or carried into the site by prehistoric people. If bison and perhaps caribou were present on the Olympic Peninsula during at least some of the last 15,000 years, then clearly the floral and geological habitats of the Olympics developed *with* those species prior to 1925. The significance of this observation is evident: Edward Schreiner, who has studied the impacts of goats on alpine plants, states that the ecosystem of the Olympics "has evolved for thousands and thousands of years" (quoted in Luoma 1993:41).

The remains of bison and caribou, of course, do not signify that mountain goats were also present; but, like the historic loss of white-tailed deer, they underscore the fact that lists of modern mammal species found in an area are not necessarily accurate reflections of past mammalian faunas. Houston and Schreiner (1993, 1994a) do not mention these fossils and thus are not considering all relevant details of the historical biogeography of the Olympic Peninsula. To proclaim, as Houston and Schreiner (1994b:11, 1995:207) do, that "national parks were not established to attempt to recreate late Pleistocene or early Holocene biotas" does not invalidate the significance of this observation for the simple reason that the development of the biota has taken place over millennia.

EVOLUTIONARY TIME DEPTH, ENDEMICS, AND GENETICS

Houston and Schreiner (1993, 1994a) write: "We are primarily concerned with taxa that are endemic to the Olympic Peninsula or to the Peninsula and Vancouver Island." What today are considered endemic forms might have evolved only in the limited geographic area where they are presently found or might have evolved in a larger area but survived only in the limited area. Thus simply documenting the occurrence of endemics tells us little about the biogeographic history of those forms. As James Brown and Arthur Gibson (1983:228) note, "investigators search for and hope to find satisfying, simplistic explanations for the origins of endemics, but what they expect to find eventually is a complex picture of how endemics evolved in time and space, to be explained in part by past geologic events and in part in terms of ecological processes."

Houston and Schreiner (1993, 1994a) review some of the geological and ecological events that no doubt influenced the modern biogeographic patterns and endemism they describe. They note that the Wisconsin-aged Cordilleran ice sheet partially overran the northern and eastern edges of the Olympic Peninsula and filled what is now Puget Sound. The Puget Lobe of the ice sheet drained west from its southwestern corner first into the Chehalis River and eventually into Grays Harbor. Once the Cordilleran ice sheet began to melt, the Puget Lobe retreated rapidly and was essentially gone from Puget Sound by 13,000 years ago. During the time it was present, Houston and Schreiner (1993:8; 1994a:37) indicate that "although still 'peninsular' during full glacial times, the Olympic area *may have been* more 'insular' in character, particularly with respect to the dispersal of terrestrial organisms" (emphasis added).

Thus, Houston and Schreiner suspect, but do not know, that the Olympic Mountains were much like a biogeographic "landbridge island": biogeographically connected to the mainland of the Cascade Mountains prior to the Wisconsin period of glacial advance, biogeographically isolated from that mainland during the period of maximum glaciation, and subsequently biogeographically reconnected, if only in a relative sense, when the ice sheet melted. Given our present understanding of the expansion and retreat of the Puget Lobe (Easterbrook 1992), it seems that the landbridge island envisioned by Houston and Schreiner (1993, 1994a) was formed no fewer than six times during the Pleistocene. If their beliefs are correct, then there were at least three times in the past 200,000 years when the insular Olympics and the mainland Cascades were isolated from one another, but they were also connected to each other three times during those years (including the present connection of the last 10,000 years or so).

Houston and Schreiner (1993, 1994a) are here subscribing to insular biogeographic theory, without using the label. This theory has been used with some success to explain the distribution patterns of animals within wildlife-management contexts (e.g., Glenn and Nudds 1989; Newmark 1987; Patterson 1987). Houston and Schreiner are proposing that the insular character of the Olympic Mountains, which varied through time, precluded the colonization of the Olympics by various nonvolant mammalian (and plant) species. They suggest that the Olympics would have been more isolated during periods of maximum glacial extent than during periods when glaciers were small or nonexistent. This is in direct contradiction to Harington's (1971) hypothesis that it was during periods of maximum glacial extent that mountain goats dispersed to new areas.

The degree to which such landbridge islands are in fact isolated is difficult to measure. For example, analysts have sometimes "presumed a greater degree of isolation for habitat remnants than the data support; depending on the [habitat] matrix surrounding a particular [habitat] remnant, some species may not be isolated at all, while others may, in fact, be trapped on what is for them the equivalent of an oceanic island" (Doak and Mills 1994:616). Houston and Schreiner (1993, 1994a) do not explicitly relate any data they present to their beliefs about the temporally fluctuating insular character of the Olympics (which are no doubt correct to some degree). Most biogeographers would agree that at a minimum such data should involve not just the habitats in the areas of concern, but also the dispersal capabilities and ecological predilections of the taxa involved. Biogeographers also indicate that "the nature and timing of past long-distance movements must be inferred from indirect evidence, such as the distributions of living *and fossil* forms" (Brown and Gibson 1983:196; emphasis added). Houston and Schreiner make no serious attempt to develop such a line of reasoning. Their implication that endemic forms and missing mammals constitute relevant data is fallacious because we simply do not know why some forms are endemic and others are missing, as they admit.

Houston and Schreiner (1993:9; 1994a:40) write: "We do not know if any of the endemic mammals of the Olympic Peninsula persisted locally during the late Wisconsin glaciation, or if all the differentiation occurred during post-Pleistocene time—which certainly seems plausible." They are in part concerned with explaining the five endemic forms of mammals (one species and four subspecies) found only on the Olympic Peninsula (see table 7.3).

One might question the validity of some of the subspecies designations; as noted by Murray Johnson and Terry Yates (1980:4), the snow mole "may well be judged a full species when additional information ... becomes available." Over a decade after their research, relevant additional data are still not available. In fact, the snow mole is still only known to occur in the single location where it was originally identified (Carraway et al. 1993), suggesting that we may not be dealing with a

TABLE 7.3.
Endemic Mammal Forms of the Olympic Peninsula

Common name	Taxonomic name	Reference
Olympic marmot	*Marmota olympus*	Hall 1981
Olympic yellow-pine chipmunk	*Eutamias amoenus caurinus*	Hall 1981
Olympic snow mole	*Scapanus townsendii olympicus*	Johnson and Yates 1980
Olympic Mazama pocket gopher	*Thomomys mazama melanops*	Hall 1981
Olympic ermine	*Mustela erminea olympica*	Hall 1981

Sources: From Houston and Schreiner 1993, table 4.1; Houston and Schreiner 1994a, table 1; Olympic National Park 1995, table 2.

unique subspecies at all. Specimens of snow mole attributed to *Scapanus townsendii olympicus* are presently known to differ from other populations of the species simply by being smaller and differing in behavior. There is some debate regarding the utility and validity of size and behavior as criteria for distinguishing separate taxa. The taxonomic status of mammal populations in the Olympics has changed over the years, as Scheffer's (1946:5) fifty-year-old comment shows: "None of the mammals of the [Olympic] peninsula ... differs more than subspecifically from its nearest relative in the Cascades." He listed the Olympic marmot as *Marmota caligata olympus* (Scheffer 1946:79), while some of his contemporaries listed it under its 1990s name, *Marmota olympus* (Dalquest 1948:267).

Subspecies designations are constantly being debated and refined as new data are gathered and new analyses are performed (e.g., Johnson 1968; Johnson and Benson 1960; Johnson and Ostenson 1959). Houston and Schreiner (1993, 1994a) do not refer to potentially relevant fossil data (e.g., Burns 1987) or to relevant evolutionary theory (e.g., Mayr 1993 and references therein) that might hint at how, why, and when these endemic mammals evolved.

Based on the similarities and differences in their genetics, the "endemic" (Hutchins and Stevens 1981:61; Houston and Schreiner 1993, 1994a) and "native" (Scheffer 1993a:918) Olympic marmot (*Marmota olympus*) and the hoary marmot (*M. caligata*) of the Cascade Mountains shared an ancestor sometime during the Pleistocene (Hoffman and Nadler 1968; Hoffman et al. 1979; see also Johnson 1968). Evolutionary divergence through time probably resulted from isolation of two populations and a lack of subsequent biogeographic dispersal from one mountain range to the other across the Puget Lowlands, a lack of subsequent genetic exchange between the two populations, or both. Unfortunately, we do not know how much time would have been necessary for the measured genetic differences to have evolved, but this does not mean this information is not useful.

Today there are five species of mammals with Olympic populations that are geographically isolated from populations in the Cascades: the northern water shrew (*Sorex palustris*), the heather vole (*Phenacomys intermedius*), the yellow-pine chipmunk, the fisher (*Martes pennanti*), and the marten (*Martes americana*). According to Hall's (1981) taxonomy, of these five only the yellow-pine chipmunk (listed by Houston and Schreiner) is represented by one subspecies in the Olympics and a different subspecies in the western Cascades (see table 7.3); the other four taxa are represented by the same subspecies in both areas. If we assume that the rate of genetic divergence of isolated populations is more or less the same across all mammalian families, then the five species with disjunct distributions and the genetics of the marmots indicate that the marmot was the first taxon to become genetically isolated, the yellow-pine chipmunk was second, and the other four species were last. These taxa may have immigrated to the Olympics in this order as well, although this is not at all assured. The subspecies of pocket gopher (*Thomomys mazama melanops*) and Olympic snow mole that are apparently unique to the Olympics (see table 7.3) also indicate the potential for dispersal and subsequent genetic isolation events later than that apparently experienced by *Marmota*.

In any case, if the rate of genetic differentiation is approximately the same across the species indicated then the order of genetic isolation of the populations of these species is approximately correct. This suggests (but does not demonstrate) that different mammal species stopped moving back and forth (or at least stopped exchanging genetic material) between the insular Olympics and the mainland Cascades at different times in the past. In turn, this implies that the dynamic glacial history of western Washington may well have hindered genetic exchange for different species at different times. Although this does not mean that mountain goats lived in the Olympics prior to 1925, it does indicate that the biogeographic history of the insular Olympics and the mainland Cascades was quite dynamic, with multiple dispersal, genetic exchange, or isolation events.

Have the genetics of the "exotic" goats in the Olympics been studied (e.g., Cronin 1993 and references therein)? This is a particularly critical question because a seventy-year-old population of mountain goats that began with twelve individuals should display a different kind of genetic variation than a long-resident large population (say, a hundred or more individuals over 10,000+ years). This expectation is based on what is known as the "founder effect," which involves the removal of a small sample of individuals from a large population and placing that sample in a (usually geographically isolated) position so that they are genetically isolated from the parent population. Because the individuals in the isolated population contain a small sample of the total range of the genetic variation found in the parent population, one can sometimes detect a founder effect in the genetics of a small founding population.

Inbreeding among the members of the founding population will exacerbate the loss of genetic variability. Typically, the longer a founding population is isolated from its parent population, the greater the differences will be; thus two closely related populations may in fact be of different species. Contracting relictual populations that result from climatic change or populations in refugia are analogous situations where the small populations undergo a genetic bottleneck (Mayr 1993 with references). Such bottlenecks can be critical factors for relictual populations of slowly reproducing ungulates such as mountain goats (see Mills and Smouse 1994 for a theoretical discussion).

This potential way to answer the question of whether the Olympic goats are native or exotic has been discussed in print in only a preliminary fashion. A single study of Olympic-goat genetics mentioned in Stevens (1983) indicates minimal individual variation in three proteins. This does not conclusively demonstrate that mountain goats were never native to the peninsula, but to one biologist "it strongly suggests that genetic variation has been lost as a result of founder effects and inbreeding" (Hutchins 1995:1325). Note that Hutchins uses the term "founder effect" (rather than the more neutral "genetic bottleneck"), implying that the lack of genetic variation derives from the dozen goats transplanted to the Olympics. The possibility that the lack of variation might derive from a low number of goats in the Olympics at an earlier time is not considered.

Finally, Houston and Schreiner (1993, 1994a) do not mention studies of endemism on other islands in the Pacific Northwest (e.g., Moodie and Reimchen 1976). This is curious because they do consider the faunas of those other islands in other contexts.

INSULAR FAUNAS AND MISSING MAMMALS

Houston and Schreiner (1993:9; 1994a:37) suggest that "a comparison of the biogeography of the Olympic Peninsula to that of the landbridge islands [Vancouver Island, the Queen Charlotte Islands, and the Alexander Archipelago of southeastern Alaska] is instructive": "In common with the Olympic Peninsula, all the landbridge islands have impoverished faunas compared to the adjacent mainland" (Houston and Schreiner 1993:12; 1994a:43).

> Mammals "missing" from the landbridge island faunas encompass species that were absent historically from the Olympics as well, including the mountain goat. As with the Olympics, this means either: (1) the species failed to colonize [these islands in the past] because of environmental barriers, or (2) earlier populations [on the islands] met extinction. That unoccupied habitat exists for additional species is suggested by the success of deliberate introductions. For example, six mammal species have been introduced successfully to the Queen Charlottes and three to Vancouver Island. The mountain goat has been introduced to Baranof and Chighagof Islands within the Alexander Archipelago, and to Kodiak Island, and has become established on at least Baranof and Kodiak, in addition to the Olympic Mountains. (Houston and Schreiner 1993:12–13; 1994a:43–44)

The implications of this statement are seemingly clear: the fact that exotic species can be successfully introduced implies that those species were never there, which in turn strengthens the notion that mountain goats were never in the Olympics prior to 1925. Similarly, because the islands that Houston and Schreiner consider share many mammalian taxa and are missing many of the mammals that are missing from the Olympic Mountains (see table 7.2), they favor the notion that the missing mammals probably never colonized any of the islands or the Olympics. The various islands (including the insular Olympics) share no "obvious characteristics or relationships … that would predispose [all of the island] populations to extinction" (Houston and Schreiner 1993:14; 1994a:45). One of the nine reviewers commissioned by Olympic National Park to review the Houston and Schreiner document found this argument to be "compelling."

There are three critical points here. First, do successful modern colonizations indicate that those new colonizing species were in fact never present or that "unoccupied habitats exist"? Second, are the lists of mammals missing from the various islands really so similar as to provide compelling evidence for the lack of colonization hypothesis? Third, should we expect some predisposition toward extinction among the mammals on the islands?

MODERN COLONIZATIONS

Houston and Schreiner (1993:12–13; 1994a:43) indicate that six mammalian species are known to have colonized the Queen Charlotte Islands and three mammalian species to have colonized Vancouver Island during the twentieth century. The colonizing taxa they list are given in table 7.4, along with the approximate dates of colonization (typically via artificial introduction). Houston and Schreiner (1993, 1994a) did not indicate these dates, but they are found in the literature (Carl and Guiguet 1972; Nagorsen 1990). I could find no date for the introduction of Townsend's chipmunk, but I reasoned that because A. W. F. Banfield (1974) does not mention that event it must have occurred after 1974 but prior to 1980, when S. E. Woods (1980) reports the chipmunk's presence.

The time depth for the successful colonizations is not extensive. The average duration of the introduced populations (subtracting the year of introduction from 1995) is 57.7 years (with a standard deviation of 22 years). The fact that these species have, on average, been present about 58 years does not mean that they will still be present 1,000 or even 100 years from now. These colonizations indicate nothing about the potential presence of mountain goats in the Olympics prior to 1925 because they say nothing about the long-term (geologically speaking) survival of any of these species in any of these areas.

Houston and Schreiner fail to list the red fox, a species that would strengthen their contention that mountain goats, along with various other alpine and subalpine taxa, were never present in the Olympics. Banfield (1974) lists the red fox as introduced, but Ian Cowan and Charles Guiguet (1965) do not; Houston and Schreiner cite Cowan and Guiguet but not Banfield. They also fail to list Townsend's chipmunk as introduced to Vancouver Island. Woods (1980), in a work not cited by Houston and Schreiner (1993, 1994a), indicates that this species is present on the island but does not report it as introduced. Hall (1981:351), who is cited by Houston and Schreiner, reports that it was introduced. Thus the

TABLE 7.4.
Recent Mammalian Colonists of Vancouver Island and the Queen Charlotte Islands (date of colonization in parentheses)

Queen Charlotte Islands	Vancouver Island
Tamiasciurus hudsonicus (1950) red squirrel	*Sylvilagus floridanus* (1964–65) eastern cottontail
Castor canadensis (1936[a] and 1950) beaver	*Sciurus carolinensis* (1914) eastern gray squirrel
Ondatra zibethicus (1924–25) muskrat	*Ondatra zibethicus* (1922) muskrat
Odocoileus hemionus [*sitkensis*] (1900 and 1912) black-tailed deer	*Vulpes vulpes*[b] (1940s) red fox
Cervus elaphus (1931–32) elk or wapiti	*Eutamias townsendii*[c] (1975?) Townsend's chipmunk
Procyon lotor (1940) raccoon	*Oreamnos americanus*[d] (1924) mountain goat

Source: Carl and Guiguet 1972; Nagorsen 1990.
[a] This introduction did not result in successful colonization (Carl and Guiguet 1972:89).
[b] This species was not listed by Houston and Schreiner (1993, 1994a). Banfield (1974:300) indicates it has been introduced to Vancouver Island.
[c] This species was not listed by Houston and Schreiner (1993, 1994a). Hall (1981:351) indicates it has been introduced to Vancouver Island. Woods (1980:132) does not indicate that it has been introduced, merely noting that it is found on the "southern portion of Vancouver Island."
[d] This species is not listed by Houston and Schreiner (1993, 1994a). Carl and Guiguet (1972:92) indicate it was introduced in 1924, but was last seen in 1936.

historical biogeographic implications of the red fox and Townsend's chipmunk are different. The red fox apparently was not native to either the Olympics or Vancouver Island during the historical period. Townsend's chipmunk is native to the Olympics at this time but not, apparently, to Vancouver Island. Perhaps this is why Houston and Schreiner do not list this species.

Houston and Schreiner (1993, 1994a) fail to report that two unsuccessful attempts were made to introduce goats to Chichagof Island (in 1953 and in 1955), that the successful Baranof Island transplant took place in 1923, and that a so far successful introduction to Revillagigedo Island took place in 1983 (MacDonald and Cook 1996:587). They also fail to mention that mountain goats were introduced to Vancouver Island in

1924, survived there just over a decade, and were last seen in 1936 (Carl and Guiguet 1972:92). Unsuccessful introductions, by their reasoning, suggest that habitats in these places (at least during the twentieth century) were inappropriate for the goats. The recently reported recovery of 12,200-year-old mountain goat remains from an alpine cave on Vancouver Island (Nagorsen et al. 1996:42) also suggests that historic habitats may be inappropriate for the modern survival of goats.

Houston and Schreiner (1993, 1994a) also fail to note that it took two attempts (an unsuccessful one in 1936 and another one in 1950) to establish beaver on the Queen Charlotte Islands (Carl and Guiguet 1972:89). This, too, potentially weakens the significance of their observation that the beaver was introduced successfully. They do not mention that Edward Hill (1982:256–57) states that beaver were native to the Queen Charlotte Islands (unfortunately, Hill provides no evidence to substantiate his statement).

It is unclear why Houston and Schreiner (1993, 1994a) list the eastern cottontail and eastern gray squirrel as successful recent colonists of Vancouver Island. These taxa today are found no closer than some 900 miles east of the Olympic Peninsula and the island (Hall 1981); it is unlikely (although not impossible) that they would have colonized the indicated west-coast areas naturally. I thus believe that these two taxa are irrelevant to any argument that various mammalian species found on the (Cascade Range) mainland should occur on the islands of the eastern Pacific Rim. By including these species Houston and Schreiner effectively lengthened the list of potentially vacant habitats on Vancouver Island by 67%, which, of course, strengthens their argument.

I could make the same argument that there are vacant ecological habitats in the Puget Sound area, as evidenced by the introduction and thus far successful colonization of that area by the opossum (*Didelphis marsupialis*). This species is not native to the western states and first appeared in northeastern Oregon about 1910 and in Washington about thirty years later (Booth 1947:20–21; Dalquest 1948:121). In the context of Houston and Schreiner's discussion this implies that this species should have been present naturally in Washington, Oregon, and California simply because it can survive there now. The biogeographic problem of why it was not found in those states naturally is not yet resolved, and the same problem exists with the introduced mammals listed in table 7.4.

One can also question the validity of the implication of Houston and Schreiner's (1993, 1994a) statement that vacant habitats exist today and thus must have existed in the past, because humans have caused modifications to habitats in the twentieth century. Would the species listed in table 7.4 have been successful colonists had they been introduced to the islands in, say, the seventeenth century? The significance of this observation is clear in the case of the coyote, which colonized the Olympic Peninsula as a result of land clearing (Scheffer 1946:58–59); early in this century, it was "invading western Washington as lumbering and clearing proceed; already reported [but not previously, in the] eastern portion of the Olympic Peninsula" (Taylor and Shaw 1929:13). Other examples exist as well.

Sea otters (*Enhyrda lutris*) were once found all along the Pacific coast. By the middle of the nineteenth century they were no longer found on the British Columbia, Washington, Oregon, or northern California coasts due to overexploitation by commercial hunters. Despite artificial introductions of new individuals to these areas in the 1970s, this species has failed to reestablish itself in southern latitudes (see Lyman 1988b and references therein). According to Houston and Schreiner's beliefs, sea otters should have had no trouble becoming reestablished; it is very well documented that a vacant habitat and a vacant niche exist. I wonder, however, if the *same* vacant habitat existed in 1980 as existed in 1580.

Artificially introduced species that have survived in a newly colonized area indicate that the ecological requirements for survival of the organism are present (e.g., significant predators are largely absent or ineffective), but this does not mean that these requirements were always present (e.g., natural predators were always absent). Given sufficient time, habitats change. An organism may be capable of survival within a particular habitat, but it may not have survived there in the past due to changes in the habitat. The distribution of an organism at any one time depends on its environmental tolerances—an ecological factor—and also on its history of distribution and associations with other species—a historical factor. As many paleobiologists have noted, "the ecological associations of taxa were different in the past from those of today" (Vrba 1992:9).

Identifying modern vacancies in an ecosystem by recording successful introductions thus tells us nothing about the probability that one of those vacancies should have been filled in the past. For one thing, there are many reasons why an introduction may or may not be successful (Colwell 1992). For another, vacancies may simply not have existed in the past. As Houston and Schreiner (1993, 1994a) admit, environments of the Olympic Peninsula have not been stable over the last several tens of thousands of years. It is paradoxical for them to imply that habitats there were stable throughout time because vacant habitats are evident today. Thus argument based on successful modern colonists (and missing mammals) fails to convince me that mountain goats were never present in the Olympics prior to 1925.

Compelling Missing Mammals?

Granting for the sake of argument (which Houston and Schreiner do) that mammal species that appear to be missing from a location are in fact missing and have not merely been overlooked, we can ask the following question: Are the "impoverished" landbridge-island mammalian faunas that Houston and Schreiner compare really so similar as to provide a "compelling" argument that all three sets of insular faunas are depauperate as a result of failure to colonize? Houston and Schreiner (1993:12, table 4.3; 1994a:44, table 3) state that Vancouver Island today lacks nineteen species (but they list twenty-one in their table) found on the mainland, and various islands making up the Alexander Archipelago today lack eighteen species. Comparing fauna from Olympic Peninsula (never a true oceanic island) with those from Vancouver Island and the Alexander Archipelago (not necessarily landbridge islands) may be faulty. Furthermore, Vancouver Island is a single large island, whereas the Alexander Archipelago consists of multiple islands of varying size that are various distances from the mainland and from each other. Such archipelagos tend to have rather complex biogeographic histories compared to large single islands (e.g., Brown 1986 and references therein).

Houston and Schreiner chose to ignore the numerous relatively small islands located between the mainland coast of British Columbia and Vancouver Island. Including them in a more detailed comparison with the Alexander Archipelago might prove illuminating. For example, ten islands between the British Columbia mainland and the Queen Charlotte Islands and Vancouver Island have modern populations of masked shrew (Cowan and Guiguet 1965; van Zyll de Jong 1983:67), a species not present on the Queen Charlottes or Vancouver Island. The numbers of British Columbia islands on which the various "missing mammals" are found are shown in table 7.5. While a detailed analysis of these data is be-

TABLE 7.5.
Mammals Missing from Vancouver Island

Species listed by Houston and Schreiner 1993, 1994a	Cowan and Guiguet 1965	Banfield 1974	Distribution on mainland coast of British Columbia adjacent to Vancouver Island
masked shrew *Sorex cinereus*	46	10	all of coast (on 12 B.C. islands)
Pacific coast mole *Scapanus orarius*	—	31	southern ¼ of coast
pika *Ochotona princeps*	94	74	all of coast
showshoe hare *Lepus americanus*	103	84	all of coast (on 1 B.C. island)
mountain beaver *Aplodontia rufa*	—	95	southern ¼ of coast
northwestern/yellow-pine chipmunk *Eutamias amoenus*	141	104	southern ⅔ of coast (on 1 B.C. island)
northern flying squirrel *Glaucomys sabrinus*	159	146	all of coast (on 3 B.C. islands)
bushy-tailed wood rat *Neotoma cinerea*	193	177	all of coast
northern bog lemming *Synaptomys borealis*	197	190	all of coast (on 2 B.C. islands)
boreal redback vole *Clethrionomys gapperi*	204	183	northern ¾ of coast (on 6 B.C. islands)
long-tailed vole *Microtus longicaudus*	226	217	all of coast (on 5 B.C. islands)
muskrat *Ondatra zibethicus*	231	200	southern ¼ of coast
porcupine *Erethizon dorsatum*	—	236	all of coast
coyote *Canis latrans*	279	288	southern ¼ of coast (on 1 B.C. island)

Species listed by Houston and Schreiner 1993, 1994a	Cowan and Guiguet 1965	Banfield 1974	Distribution on mainland coast of British Columbia adjacent to Vancouver Island
red fox *Vulpes vulpes*	288	300	southern ¼ of coast
grizzly bear *Ursus arctos*	—	310	all of coast
fisher *Martes pennanti*	308	319	all of coast
spotted skunk *Spilogale gracilis*	—	338	southern ⅔ of coast
stripped skunk *Mephitis mephitis*	—	340	southern ¼ of coast (on 1 B.C. island)
bobcat *Lynx rufus*	341	353	southern ½ of coast
mountain goat *Oreamnos americanus*	391	410	all of coast
Species not listed by Houston and Schreiner (1993, 1994a)			
Bendire's shrew *Sorex bendirii*	55	15	southern ¼ of coast
Trowbridge shrew *Sorex trowbridgii*	—	19	southern ¼ of coast
American shrew-mole *Neurotrichus gibbsii*	—	28	southern ¼ of coast
Townsend's mole *Scapanus townsendii*	—	29	southern ¼ of coast
Townsend's chipmunk[a] *Eutamias townsendii*	144	105	southern ¼ of coast
western red-backed vole *Clethrionomys occidentalis*	204	179	southern ¼ of coast
heather vole *Phenacomys intermedius*	201	192	all of coast

Species not listed by Houston and Schreiner (1993, 1994a)			
creeping vole *Microtus oregoni*	215	221	southern ½ of coast
Pacific jumping mouse *Zapus trinotatus*	243	225	southern ½ of coast (on 1 B.C. island)
long-tailed weasel *Mustela frenata*	318	324	southern ¼ of coast

Note: The numbers in the two middle columns are the page numbers for the range map for the indicated species; "—" indicates that no map was included (only verbal descriptions of the range were provided).
[a]This species has been recently introduced to the island (see table 7.4).

yond the scope of what Houston and Schreiner are attempting to do, inclusion of those island records would certainly make the comparison more reasonable in terms of modern biogeographic theory.

Houston and Schreiner's (1993, 1994a) list indicates that the hoary marmot, the northern redback vole (*Clethrionomys rutilus*), the muskrat, the jumping mouse (*Zapus hudsonius*), the porcupine, the wolverine (*Gulo luscus*), the moose (*Alces alces*), and the mountain goat are not found on *any* of the islands of the Alexander Archipelago. Reviewing the literature they cite (Cowan and Guiguet 1965; Manville and Young 1965) and an additional reference (Swarth 1936) that they did not include, I find that all of these species occur on one or more of the islands of the archipelago (see also MacDonald and Cook 1996). Omitting these incorrectly listed species, the Alexander Archipelago as a whole is not impoverished by eighteen mammal species but by only six species: the northern water shrew, the pika, the snowshoe or varying hare (*Lepus americanus*), the coyote, the red fox, and the lynx (*Lynx canadensis*).

Houston and Schreiner (1993:table 4.3; 1994a:table 3) also indicate that several species they list as "absent from the Alexander Archipelago" are, in fact, found on some of the islands. But they do not offer any reason for including these species in the list of mammals missing from the archipelago. In a later article, Houston (1995:239) reiterates that some mammals are missing from one or more islands of the archipelago and indicates that what I perceive to be errors in listing actually "result from differing interpretations." I believe, however, that it is not a matter of interpretation but a result of what we each sought to show. Houston was concerned with showing that these missing mammals were not very successful at colonizing the islands of the archipelago. I was (Lyman 1995a) and am concerned with showing that they could have colonized such islands, later becoming extinct on some of them.

One of the four species that Houston and Schreiner note as present on one or a few of the islands of the Alexander Archipelago, the northern bog lemming, is also absent from both Vancouver Island and the Olympics; another, the northern flying squirrel, is absent from Vancouver Island, though it is present on Quadra Island (Cowan and Guiguet 1965:162). Houston and Schreiner never specify the rules they used to construct the lists of missing mammals. At any rate, including these species makes their lists of mammals "absent from" the various insular faunas appear more similar than they really are. They also do not list the record of a mountain goat on Wrangell Island in the Alexander Archipelago (Klein 1965; Manville and Young 1965).

A seemingly convincing argument that mountain goats were not present in the Olympics prior to 1925 because those mountains are insular and thus multiple species of mammals could not colonize them can be based on cases of insular mountains that have depauperate mammalian faunas, all lacking more or less the same species. I suspect that this is why one reviewer found Houston and Schreiner's argument to be so "compelling." The presented lists are quite similar, but the published lists are not accurate.

The data concerning mammals missing from Vancouver Island (Houston and Schreiner 1993:table 4.3; 1994a:table 3) appear to be quite inaccurate. These data are summarized in table 7.5, along with some additional information. I consulted the same reference that Houston and Schreiner did (Cowan and Guiguet 1965), plus two more recent synopses of British Columbia mammals (Banfield 1974; Nagorsen 1990). At least ten additional species not included in Houston and Schreiner's list of mammals missing from Vancouver Island (but all listed in the single reference they cite) should be included if the criteria used to select those species are consistently applied. I reasoned that mammalian species that only occur on a fraction of the British Columbia mainland coast adjacent to Vancouver Island would probably have less chance to colonize that island than species that occur along the entire mainland coastline adjacent to the island. Houston and Schreiner apparently used such a criterion when they constructed their list of mammals missing from Vancouver Island (Houston 1995), but it was not consistently applied.

Six of the taxa listed by Houston and Schreiner are today found on only the southern one-fourth of the mainland coastline adjacent to Vancouver Island (table 7.5). Seven other taxa that they do not list as missing from Vancouver Island but that are in fact missing are also found along only the southern one-fourth of the mainland coast. Should species such as these with rather limited mainland distributions be eliminated from the list? If so, Houston and Schreiner's list would be reduced from twenty-one species to eighteen species if the three additional species not included that in fact occur on more than one-fourth of the mainland coast are added. But if all species that occupy any fraction of the mainland coast adjacent to Vancouver Island are included, then the list of mammals missing from the island is not nineteen species as stated in their text or twenty-one as indicated in their table 3; it is the thirty-one species listed in table 7.5. There is a simple way to show that Houston and Schreiner's lists of mammals missing from the Olympics, Vancouver Island, and the Alexander Archipelago look more similar than the lists I have constructed.

Biologists have developed a variety of indices to measure the similarity of two lists of species (see Cheetham and Hazel 1969 and Raup and Crick 1979 for overviews). Some indexes emphasize similarities whereas others emphasize differences. Because I wanted a conservative measure of the differences between the lists, I chose an index that emphasizes similarities. The Sorenson Index is calculated with the formula

$$[2C/(A + B)]\ 100 = S$$

where C is the number of species in common on the two compared species lists, A is the total number of species in one list, B is the total number of species in the other, and S is the index value. The Sorenson Index, S, is similar to a measure of the percentage of all the species found in two lists that are shared by the two lists. We can measure the change in the similarity of the various lists of mammal species missing from the Olympics, Vancouver Island, and the Alexander Archipelago by comparing S values calculated using Houston and Schreiner's lists with those calculated using my lists for each of the three insular faunas.

Data used in the calculations are presented in tables 7.2 and 7.5 and in the discussion of the mammals missing from the Alexander Archipelago above. According to Houston and Schreiner's lists, the Olympics are missing twelve species, Vancouver Island twenty-one species, and the Alexander Archipelago eighteen species. The number of missing species shared by each possible pair is nine for Vancouver Island and the Alexander Archipelago, seven for the Alexander Archipelago and the Olympics, and eight for Vancouver Island and the Olympics. The *S* values for each possible pair of areas are Vancouver Island/Alexander Archipelago 46, Vancouver Island/Olympic Mountains 48, and Olympic Mountains/Alexander Archipelago 47. The *S* values indicate that about half of the missing species are found in each pair of lists; thus the lists of missing mammals produced by Houston and Schreiner appear somewhat similar and might provide "compelling" evidence that these mammals probably never colonized any of these insular locations.

If we use the lists of missing mammals I constructed, Vancouver Island has thirty-one missing mammal species (table 7.5), the Alexander Archipelago has six (only species missing from all of the islands of the archipelago are included), and the Olympics eleven (the northern bog lemming is omitted for reasons discussed above). Vancouver Island and the Alexander Archipelago share four missing species, the Alexander Archipelago and the Olympics share three, and Vancouver Island and the Olympics share seven. The *S* values calculated using these numbers are Alexander Archipelago/Vancouver Island 22, Olympic Mountains/Vancouver Island 33, and Olympic Mountains/Alexander Archipelago 35. These *S* values indicate that the lists of missing mammals for the three insular faunas are not very similar; the greatest proportion of missing mammals of any one pair is only 35%. The "compelling" evidence tends to disappear when my lists of missing mammals are used.

I offered to describe these differences in letters dated May 19, 1994, and August 25, 1994, to Park Superintendent Finnerty. While I received responses to these letters, my offer was ignored. The monograph in which Houston and Schreiner (1994a) is found was shipped to the Denver distribution center on January 13, 1995 (NPS publications coordinator, Denver, personal communication to Lyman, January 20, 1995). It was not too late to make corrections to the manuscript in May 1994. But by that time Houston (March 9, 1995, letter to Lyman) had decided that I was "an unlikely candidate to provide an impartial review of the mountain-goat issue."

In response to my observations (Lyman 1995a) that his data might contain errors, Houston (1995) indicated that he was attempting to omit species "whose current ranges in British Columbia were limited to the southwestern tip of the mainland, i.e., their ranges did not extend far north along the mainland coast parallel to Vancouver Island," and species whose distributions were more to the south. He concluded: "In retrospect, the 10 species could be included on a list of mammals considered to be absent from Vancouver Island." Houston (1995) thus admitted that he might have made mistakes and also clarified the rules he used to derive the list of missing mammals for Vancouver Island (none of the twenty-three people who reviewed Houston and Schreiner 1993 prior to publication apparently noticed this lack of rules). Following Dalquest's (1948:52) lead, Houston originally omitted seven species with relatively southern distributions of the ten species that I suggest should be considered missing from Vancouver Island. This was not made clear initially. Even given this clarification, the lists he used may not be the correct ones.

PREDISPOSITION TO EXTINCTION OR FAILURE TO COLONIZE

Houston and Schreiner (1993:14; 1994a:45) offer the following conclusion:

> the impoverished nature of the terrestrial mammal fauna of the Olympics is to be expected. We do not know relative contributions to the present fauna of extinction versus the failure to colonize. There are no obvious characteristics or relationships in common among species absent from the Peninsula and the landbridge islands that would similarly predispose populations to extinction (e.g., pika, bog lemming, porcupine, red fox, mountain goat). Densities, demography, and ecological requirements differ widely among the five species. Failure to colonize might be as plausible an explanation as extinction in accounting for the absence of the four species (porcupine excepted) that occupy subalpine and alpine habitat [of the Olympics]; extensive bodies of water and forest may have represented effective barriers to colonization.

Houston and Schreiner suggest that it does not seem likely that these five species colonized the Olympics at some time in the past but became locally extinct sometime before the historic period because four species they list (porcupine omitted) are not found on Vancouver Island, the Queen Charlotte Islands, or the islands of the Alexander Archipelago. However, as I have noted, many of their data on modern mammal distributions seem to be inaccurate; thus the implication of their analysis is compromised.

Furthermore, Houston and Schreiner ignore relevant fossil data. Paleontological research in the Alexander Archipelago over the last several years has produced remains of caribou, red fox, grizzly bear, and other taxa on Prince of Wales Island (Heaton 1994, 1995, 1996, 1997; Heaton and Grady 1992a, 1992b, 1993; Heaton et al. 1996; see also Busch 1994). The grizzly bear is not historically known to occur on Prince of Wales Island. The red fox and caribou are not historically reported on any island in the archipelago. Late Pleistocene remains of mountain goats have been found on Vancouver Island, where this species is not historically known (Nagorsen et al. 1996). These few fossils (little paleontological work has been done on the Alexander Archipelago or in alpine habitats on Vancouver Island) underscore the fact that the lists of missing mammals in tables 7.2 and 7.5 tell us very little about the historic biogeography of insular locations along the Northwest Coast. The fossils indicate that the red fox and mountain goat should be excluded from Houston and Schreiner's parenthetical list of mammals cited above.

Houston and Schreiner use the patterned absence of four species (pika, northern bog lemming, red fox, and mountain goat) from all of the islands in conjunction with the lack of some environmental factor prompting a predisposition to extinction to suggest that these species never colonized any of the islands. I perceive three problems in their position. First, it is clear that the choice of these particular species is a function of their lists of missing mammals. These lists are potentially inaccurate because they presume that these species never colonized the islands. Second, if these species did colonize the islands, should we expect that there was some environmental factor that predisposed them to become locally extinct on all islands? That is, should we expect patterned extinction of isolated populations of the same species in different montane islandlike habitats? Third, if we turn the previous question around, is there some environmental factor that should have predisposed these species not to colonize any of these island habitats?

If the list of species is inaccurate, the other questions become irrelevant because they presume that the list of species is accurate across all islands considered. We can, however, briefly consider the two other points from a more conceptual perspective, without reference to any particular list of species. Many studies of insular biogeography indicate that it is extremely difficult to predict which taxa will become extinct and in which insular location(s) they will become extinct because so many variables can be involved in driving an isolated population to extinction (Brown

1986; Cutler 1991; and references therein). A patterned extinction of, for example, the pika, bog lemming, red fox, and mountain goat across the Olympic Mountains, Vancouver Island, Queen Charlotte Islands, and Alexander Archipelago may seem improbable, but it is not impossible. It seems more parsimonious to conclude that this set of taxa never colonized these four insular locations. This is no doubt why Houston and Schreiner prefer this conclusion.

Because colonization of an area depends on many variables (e.g., access to the new area; ecological tolerances on the part of the colonists; successful reproduction; sufficient genetic variation to avoid inbreeding, genetic stagnation, and eventual extirpation), it seems probable that a complete set of taxa will not colonize precisely the same set of insular areas. The historical contingencies of extinction in an insular area, however, require colonization prior to local extirpation. Thus one more step is necessary for the predisposition-to-extinction possibility than for the predisposition-to-colonization possibility. This additional step alone might lead us to align with Houston and Schreiner, who seem to favor the failure to colonize hypothesis rather than the colonization followed by extinction hypothesis. Such alignment does not, however, account for the fossil record, a critical source of evidence that casts doubt on the failure to colonize hypothesis.

HYPOTHESES, IDEAS, DATA, AND THE GROWTH OF SCIENTIFIC KNOWLEDGE

One would expect to find at least a brief review of previous hypotheses regarding the biogeography of the biota of the Pacific Northwest in Houston and Schreiner's (1993, 1994a) report, as well as a discussion of how and why particular data were chosen for analysis while other data were not used. I find neither a review of previous ideas nor any explicit data screening.

Houston and Schreiner cite Johnson and Yates (1980), but they do not refer to the following statement: "A number of species of small plants are endemic to the Olympic Mountains. Several other species occurring in the drier northeastern sector (including Hurricane Ridge) are characteristic of the arid region of eastern Washington. *The former may have developed during isolation, whereas the latter species may have reached the area during drier times*" (Johnson and Yates 1980:2; emphasis added). Here is a rather concise historical explanation for the endemism of the plants of the Olympic Mountains documented by Houston and Schreiner (1993, 1994a). This statement could be structured as a hypothesis and might be testable with fossil pollen, but Houston and Schreiner do not mention such a possibility, nor do they consider any fossil botanical data.

Houston and Schreiner use data from Hall's (1981) book on the mammals of North America, but they do not note that Hall showed the range of three of the species listed in table 7.2 as including the Olympic Peninsula. While Hall (1981:viii) stated that his biogeographic range maps sometimes represented "only the author's guesses as to occurrence" of species and subspecies, at least one other biogeographer (Newmark 1986a, 1987) found those maps to be acceptable. In fact, Newmark (e.g., 1987) is cited with favor by several individuals who argue that mountain goats and the other species listed in table 7.2 were not historically present in the Olympics (Houston and Schreiner 1993, 1994a; Scheffer 1993a). Newmark (1986a:122) suggested that grizzly bear, wolverine, and lynx were present on Olympic National Park lands and were locally extirpated prior to establishment of the park. This aspect of Newmark's work and Hall's distributional map data are not mentioned by Houston and Schreiner (1993, 1994a). To my knowledge, there are no historical records of grizzly bear adjacent to Puget Sound or on the Olympic Peninsula (Booth 1947; Dalquest 1948). The only record of this species in either of these areas consists of fossils estimated to be 11,000 to 12,000 years old recovered from Whidby Island in northern Puget Sound and only recently reported (Mustoe and Carlstad 1995). They indicate, but do not demonstrate, that grizzly bears may have been present on the Olympic Peninsula in the past.

Houston and Schreiner (1993, 1994a) do not mention published studies on the biogeographic history of the Alexander Archipelago (Klein 1965; Swarth 1936). Thus it is perhaps not surprising that there are errors in their data for that archipelago. Further, they do not consider fossil data from the Alexander Archipelago, although preliminary descriptions were available well before they completed their manuscript (e.g., Heaton and Grady 1992a, 1992b, 1993). These data provide significant evidence for the biogeographic and deglaciation history of the archipelago (e.g., Busch 1994) and underscore that we have much to learn regarding the deep history of the mammalian biogeography of the Pacific Northwest. For example, "the first Pleistocene vertebrate fossils recovered from the

Alexander Archipelago" are extremely significant because they indicate "early post-glacial colonization followed by some extirpation [of mammals]" (Heaton and Grady 1992b:32A), which is precisely the scenario Houston and Schreiner seem to disfavor.

Houston and Schreiner (1993, 1994a) also fail to mention fossil plant and geological data that are relevant to understanding the biogeographic history of the Queen Charlotte Islands (Warner et al. 1982) and the biogeographic history of plants on the Olympic Peninsula (e.g., Petersen et al. 1983). Perhaps they believe that their research is thorough even without reference to the fossil data. Superintendent Finnerty stated in an April 1, 1994, letter to Keith Gunnar that "the paleontological record is of interest, but compared to [other disciplines such as history and mammalogy], it has little relevance to learning the status of the present-day mountain goat." If this statement were true, Houston and Schreiner might ignore evidence such as Dalquest (1948) and the fossil evidence indicating that the northern bog lemming and the bighorn sheep were found prehistorically in the southern Cascades of Washington. But, as we have seen, Finnerty's statement is not true. Houston (1995) explicitly states that he considered the effects of the Pleistocene on mammal distributions during his analysis. Thus the fossil record must be relevant.

Prior to 1988, there was only one model explaining the historic biogeography of the Olympic Peninsula mammalian fauna: Merriam's, as later modified by Scheffer. It indicated that the absence of mammals from the Olympics was a result of failure to colonize; this was the dogma of the 1980s. An alternative model focusing on the mountain goat and suggesting colonization followed by extinction was published in 1988 (Lyman 1988a). Yet Houston and Schreiner (1993) do not mention it in their manuscript, though it is mentioned in the introduction to the monograph (Houston and Schreiner 1994) and in Schultz's (1994) review of the historical data. While they refer to one study of the mammalian biogeography of western parks (Newmark 1987), they do not cite several related papers (e.g., Glenn and Nudds 1989; Newmark 1986a, 1986b). These uncited papers (as well as the one they do cite) contain detailed analyses of the type that should be performed with the insular and island faunas described by Houston and Schreiner (1993, 1994a). Such analyses often provide insights into various aspects of the biogeographic history of the organisms. Similarly, an important historical biogeographic study (Clarke and Johnson 1990) of the white-tailed ptarmigan (*Lagopus leucu-*

rus), the one bird species listed by Houston and Schreiner (1993:table 4.2; 1994a:table 2) as missing from the Olympics but found in the Cascade Mountains, is not referenced by them. That study could easily have been emulated in the Olympics.

WHY WEREN'T MOUNTAIN GOATS FOUND IN THE OLYMPICS PRIOR TO 1925?

As a final comment on the historical biogeography of mountain goats in the Pacific Northwest, it is appropriate to summarize some indirect evidence (not considered by park biologists) that mountain goats and other alpine mammals presently missing from the Olympics might have arrived there prior to 1925 and to explain why I think mountain goats and these other species might be historically missing from those mountains.

Indirect Evidence for the Prehistoric Presence of Mountain Goats

Following Bruce Patterson (1984, 1987, 1990; Patterson and Atmar 1986), if the Olympic Mountains represent an insular habitat connected at some time in the past with the southern Cascade Mountains of Washington (the probable "mainland" source of many of the mammals in the Olympics), then the montane mammalian fauna of the Olympics should be a subset of the fauna of the southern Cascades. Given Dalquest's (1948) biogeographic model and limited fossil evidence, the source of the montane mammalian fauna of the southern Cascades most probably is the northern Cascades of Washington; thus the fauna of the southern Cascades should be a subset of the fauna of the northern Cascades. These three faunas should constitute a "nested subset" in which "smaller biotas contain successive subsets of the species in larger biotas" (Patterson 1990:331).

Such a series of nested subsets would suggest, but not unequivocally demonstrate, that prehistoric colonization followed by extinction was a factor contributing to the observed distribution of mammals, whereas failure to colonize perhaps played a more minor role (Patterson 1990). Patterson's (1990) research suggests that nested subset patterns are more likely to result from differential extinction than from differential colonization; thus the presence of nested subsets among Washington montane mammals would suggest that the probability that mountain goats were prehistorically present in the Olympics is greater than zero, though nested subsets would not demonstrate that they were present because a series of nested subsets could result from a failure to colonize as well as from colonization followed by extinction. Biologists are presently grappling with whether particular nested subset patterns denote selective colonization or selective extinction (e.g., Cook and Quinn 1995; Lomolino 1996). Sorting out which possibility is the more likely demands fossil data; such data have begun to clarify the historical reasons for the nested pattern of montane mammals in Nevada and contiguous areas (e.g., Grayson et al. 1996). If prehistoric faunas contain remains of mammals missing from one of the nested subsets for the area where the remains were found, then colonization followed by extinction would be the more likely alternative.

I compiled lists of nonvolant, terrestrial mammal species historically found in the northern Cascades of Washington, the southern Cascades of Washington (especially Mount Rainier), and the Olympic Peninsula from several sources (Dalquest 1948; Dice 1932; Hall 1981; Johnson and Johnson 1952; Macy 1934; Schamberger 1970; Scheffer 1946; Svihla and Svihla 1933; Taylor and Shaw 1927). To control partially for taxa that may have colonized some of these areas from more southern latitudes, five species that are today found on Mount Hood in Oregon but not in the northern Cascades of Washington were omitted from the species lists (*Sorex bendirii*, Bendire's [Pacific water] shrew; *Marmota flaviventris*, yellow-bellied marmot; *Sciurus griseus*, western gray squirrel; *Arborimus longicaudus*, red tree vole; *Microtus townsendii*, montane vole). Resulting species lists are presented in table 7.6. These lists are subsets of one another: the northern Cascades of Washington has fifty-three species, fifty of which are found in the southern Cascades of Washington. The Olympic Mountains are the most depauperate of the three faunas with forty-one species (the coyote apparently immigrated to the Olympics from areas to the east and south in about 1910 [Scheffer 1946:58] and is not included in the total), all of which are found in the southern Cascades of Washington.

It is important to note that in my analysis I treated *Marmota caligata* and *Marmota olympus* as one species. Genetic and osteometric data suggest that these two species share a common ancestor and are closely related (Hoffmann and Nadler 1968; Hoffmann et al. 1979). *Marmota olympus* is probably an endemic that evolved in the insular Olympic Mountains as a result of allopatric speciation, the divergence of a small

TABLE 7.6.
Terrestrial, Nonvolant Mammals Reported in the Olympic Peninsula, Southern Cascades of Washington, and North Cascades of Washington

Taxonomic name Common name	North Cascades	Southern Cascades	Olympic Peninsula
Sorex cinereus masked shrew	+	+	+
Sorex vagrans vagrant shrew	+	+	+
Sorex palustris[a] mountain water shrew	+	+	+
Sorex trowbridgii Trowbridge's shrew	+	+	+
Neurotrichus gibbsii shrew mole	+	+	+
Scapanus orarius coast mole	P	+As	+
Scapanus townsendii Townsend mole	P	+	+
Ochotona princeps pika	+	+	
Lepus americanus snowshoe hare	+As	+As	+As
Aplodontia rufa mountain beaver	+As	+As	+
Eutamias amoenus[a] mountain chipmunk	+	+Ag	+
Eutamias townsendii Townsend chipmunk	+	+	+
Marmota olympus/caligata[b] Olympic/hoary marmot	+Ag	+	+
Spermophilus saturatus golden-mantled ground squirrel	+	+Ag	

Taxonomic name Common name	North Cascades	Southern Cascades	Olympic Peninsula
Tamiasciurus hudsonicus red squirrel	+		
Tamiasciurus douglasii Douglas' squirrel	+	+Ag	+Ag
Glaucomys sabrinus flying squirrel	+	+As	+
Thomomys mazama/talpoides[c] Mazama/northern pocket gopher	+	+As	+
Castor canadensis beaver	+As	+As	+As
Peromyscus maniculatus deer mouse	+As	+As	+
Neotoma cinerea bushy-tailed woodrat	+As	+As	+
Clethrionomys californicus/gapperi California/red-backed mouse	+As	+	+
Phenacomys intermedius[a] heather vole	+	+	+
Microtus longicaudus long-tailed meadow mouse	+	+	+
Microtus oregoni Oregon meadow mouse/creeping vole	+	+	+
Microtus richardsoni water vole	+	+	
Ondatra zibethicus muskrat	+	+As	+As
Synaptomys borealis[d] northern bog lemming	+	prehistoric?	
Zapus trinotatus Pacific jumping mouse	+	+	+
Erethizon dorsatum[e] porcupine	+	+	+

Taxonomic name Common name	North Cascades	Southern Cascades	Olympic Peninsula
Canis latrans coyote	+Ag	+As	immigrated ca. 1910
Canis lupus gray wolf	+	+E	+As, E
Vulpes vulpes red fox	+	+	introduced
Ursus americanus black bear	+Ag	+As	+As
Ursus arctos grizzly bear	+	prehistoric?	
Procyon lotor raccoon	+	+As	+As
Martes americana[a] marten	+	+As	+As
Martes pennanti[a] fisher	+	+	+
Mustela erminea ermine	+	+	+Ag
Mustela frenata long-tailed weasel	+	+	+
Mustela vison mink	+As	+	+As
Gulo luscus wolverine	+	+	
Spilogale putorius spotted skunk	+	+	+
Mephitis mephitis striped skunk	+	+As	+As
Lutra canadensis river otter	+	+As	+As
Felis concolor cougar	+	+As	+As

Taxonomic name Common name	North Cascades	Southern Cascades	Olympic Peninsula
Lynx canadensis lynx	+Ag	+Ag	
Lynx rufus bobcat	+	+	+As
Cervus elaphus wapiti	+As	+As	+As
Odocoileus hemionus black-tailed deer	+Ag	+As	+As
Odocoileus virginianus white-tailed deer	+	+	+E
Oreamnos americanus mountain goat	+As	+As	introduced
Ovis canadensis[d] bighorn sheep	+As	+E,As	
total *N* of taxa	53	50	41[f]
N archaeofaunal genera	15	25	15[g]

Note: +, historically present; P, historically present but may be only in locations peripheral to indicated area; As, species is known from archaeological record; Ag, genus is known from archaeological record; E, locally extirpated.
[a] Olympic Mountains population disjunct from Southern Cascades population.
[b] These two species have nonoverlapping distributions, but are closely related genetically and morphologically.
[c] These two species have nonoverlapping distributions in Washington, but are closely related genetically and morphologically.
[d] These two species are historically reported only in the northern Cascades of Washington, not in the southern Cascades.
[e] This species is reported by Johnson and Johnson (1952) to be rarely present only in the southern portion of the Olympic Peninsula.
[f] Coyote not included in total.
[g] Total does not include *Bison* sp. or *Rangifer* sp., both of which have been identified in the archaeological record.

isolated daughter population from the parental population. Treating the two marmot species as separate species would reduce measures of nestedness because that is the effect of endemics on such measures (Cook and Quinn 1995).

Discovering the remains of any of the northern Cascades taxa presently missing from the southern Cascades in montane habitats in the southern Cascades and in contexts dated to the late Quaternary would lend support to the hypothesis that the latter fauna is a nested subset of the former due to prehistoric colonization and subsequent extinction. The bighorn sheep and northern bog lemming (prehistoric remains of which have been recovered from the southern Cascades of Washington, where there is no historic record of them) are prime examples. Similarly, discovery of the remains in an Olympic Peninsula site of any southern Cascades taxon that is presently missing from the Olympics (table 7.2) would increase the probability that mountain goats may well have occurred in the Olympics prior to 1925 and were (apparently) historically absent from the Olympic Peninsula as a result of extinction rather than of failure to colonize that area. Remains of bison and caribou dating to the late Pleistocene–early Holocene recovered from the Manis Mastodon Site in the northeastern corner of the Olympic Peninsula lend indirect support to the possibility of pre-1925 mountain goats in the Olympics (see the following section).

Other fossil evidence is less suggestive of the fact that mountain goats may have been present in the Olympics sometime in prehistory, but it is relevant to identifying the cause of nested subsets. Most of the species listed by Scheffer (1946) and Johnson and Johnson (1952) as not present on the Olympic Peninsula but present in the southern Cascades (table 7.2) have had relatively dynamic biogeographic histories in the Pacific Northwest. Late Quaternary (last 25,000 years) fossils of the pika (Mead 1987), the water vole (Lyman, unpublished data), the northern bog lemming (see Rensberger et al. 1984 and Rensberger and Barnosky 1993 for Washington; see Mead et al. 1992 for the western United States), the red fox (Lyman 1991, unpublished data), the mountain sheep (Lyman and Livingston 1983), the grizzly bear (Lyman 1986, unpublished data; Mustoe and Carlstad 1995), and the mountain goat (Furlong 1906; Lyman 1988a; Sinclair 1903, 1904, 1905) have been found in areas of the western United States (for all but the pika, specifically in eastern Washington state), where those species are not historically documented. Although some of these remains may have been transported significant distances from where the animals lived, it is doubtful that all of the remains were so transported.

The recently reported 12,200-year-old mountain-goat fossils from Vancouver Island (Nagorsen et al. 1996) suggest that this species had a quite different range in the past than it does at present. Further evidence that mountain goats were in different areas prehistorically than historically derives from the fossil record for the extinct Harrington's mountain goat (*Oreamnos harringtoni*) (Mead and Lawler 1994). This species apparently became extinct at the end of the Pleistocene (Mead, Martin, Euler, Long, Jull, Toolin, Donahue, and Linick 1986) but enjoyed a wide range in the southwest and southern Rocky Mountains prior to that time (Mead, O'Rourke, and Foppe 1986; Mead et al. 1987; Stock 1936).

The implication of the fossil record is clear: several montane (if not alpine or subalpine) taxa dispersed over much of the mountainous west, probably during a period of glacial advance. As Bruce Patterson and Wirt Atmar (1986:76) note, "the principal of parsimony indicates widespread dispersal during glaciopluvial periods that carried a majority of montane mammal species to a majority of mountain ranges." This, of course, does not mean that mountain goats dispersed similarly, but it does make it seem probable that mountain goats were more widely distributed in the Pacific Northwest during the last few thousand years of the Pleistocene than they were seventy-five years ago. Thus, I agree with the following statement by Patterson and Atmar (1986:76): "Having established [the above] records, it is not parsimonious to maintain that [mountain goats], now inhabiting the same microhabitats in species-rich montane faunas but absent in the available fossil record, did not similarly colonize [the Olympic Mountains]."

The evidence presented here is, at best, indirect. It does not unequivocally indicate that mountain goats were present on the Olympic Mountains prior to the 1925 transplant. The only evidence that will provide such an indication must come from the fossil record (see chapter 6; Lyman 1995b; Schalk 1993).

Why Were Goats Absent in 1925?

Biologists regularly grapple with extinction (the complete loss of a species) and extirpation (the loss of a population of a species in a particular area). Sometimes these efforts are related to the conservation problem of maintaining a minimum viable population size (e.g., Korn 1994; see chapter 3), but at other times they are on a more general level: attempts are made to correlate variables such as body size, range size, reproductive rates, and climatic stability with some measure of a population's propensity to disappear (e.g., Brown and Maurer 1989; Foley 1994; Gaston and Blackburn 1996; Korn 1994; Lande 1993; Pimm et al. 1988; Tracy and George 1992). What seems to be emerging from this research is that the probability that an isolated population will be extirpated depends on a large number of variables, including the magnitude and the timing or rapidity of climatic and habitat change (e.g., Johst and Brandl 1997; McDonald and Brown 1992).

To explain why mountain goats were not historically found in the Olympics and the Cascades of Oregon and northern California (see fig. 7.2), I suggested that "random extinction events took place during the Holocene" on the mountaintops to which goats had earlier immigrated (Lyman 1988a:20). These mountaintops would have been isolated alpine habitat islands after the Pleistocene, and their insular character would have precluded recolonization after extinction. I suggested that volcanic eruptions of some of these mountains (the Olympic Mountains are nonvolcanic), human overkill, and climatic fluctuations might have been the catalysts for extirpation.

Climatic change has recently been proposed as a cause for the extinction of mammals on mountaintops in the Pacific Northwest in a volume (Pielou 1991) cited by Houston and Schreiner (1993, 1994a), yet they do not mention climatic change as a possible reason for the prehistoric extirpation of goats in the Olympics. Climatic change could account for the absence of several of the species listed as missing in table 7.2, especially those such as the northern bog lemming and the heather vole that are clearly subalpine and alpine creatures. Further, if predation of goats by wolves (e.g., Fox and Streveler 1986), humans (e.g., Brooks 1930), and other predators kept the goat population small, climatic change may have merely administered the coup de grace to the local population.

For sake of discussion, let us assume that all of the taxa listed in table 7.2 colonized the Olympic Mountains about 15,000 years ago. Why were they not there in the early twentieth century? E.C. Pielou (1991:280–82) notes that after the end of the Pleistocene climates tended to become gradually warmer and drier, so that by about 8,000 years ago (the precise timing varies depending on geographic location) temperatures were warmer and amounts of effective precipitation were less than today. These conditions lasted until about 4,000 years ago (again, the precise timing varies from location to location), when the climate gradually shifted to modern conditions.

During this period of warm and dry climate, referred to as the hypsithermal or altithermal climatic period, "the patches of alpine tundra at high elevations dwindled" (Pielou 1991:281) as dense forests extended upslope (i.e., timberline increased in elevation). Such responses of vegetation to climate change are well documented for the historic period at

Mount Rainier (Rochefort and Peterson 1996), where fossil pollen indicate higher tree line during the hypsithermal (Dunwiddie 1986; Rochefort et al. 1994). Similar climatic and habitat changes during the hypsithermal are also indicated in southwestern British Columbia (e.g., Evans 1997). As Pielou (1991:281) notes, changes in the ranges of habitats and vegetation caused by the hypsithermal are "believed to account for the modern geographical ranges of two familiar mountain mammals, the mountain goat and the hoary marmot." During the period of maximum glacial extent, both species "ranged farther south than they do at present," and available evidence indicates "that they spread into their present ranges from the south when the Cordilleran ice sheets melted. Subsequently, as the fossil evidence shows, they died out in some places, presumably where their required habitat—alpine tundra—dwindled and disappeared in the warmth of the hypsithermal. *This may explain the present-day absence of the two species from the southern Rockies and Coast-Cascade ranges*" (Pielou 1991:282; emphasis added).

There are two intriguing aspects to Pielou's statements. First, and most relevant here, the Olympics, Vancouver Island, the Queen Charlottes, and portions of the Alexander Archipelago are all parts of the Coast Range. Thus Pielou has proposed an explanation for the historical absence of mountain goats from these modern insular areas, an explanation that Houston and Schreiner (1993, 1994a) ignore but that may apply with equal force to the other alpine and subalpine taxa listed in table 7.2. Climatic change prompting habitat diminution may well account for the apparent historic absence of mountain goats as well as the alpine/subalpine golden-mantled ground squirrel and pika from the Olympics.

Second, park biologist Bruce Moorhead (1977:2) gave a similar explanation for the ice age dispersal of mountain goats: "Fossil records indicate that early mountain goats dispersed as far south as California and Mexico, probably along the advancing edge of continental glaciers. As the climate warmed and these glaciers withdrew northward, goat distribution has progressively been restricted to high mountain retreats in the Pacific Northwest." Houston and Schreiner (1993, 1994a) ignore such statements of historical biogeography and instead focus on endemism, but they do not attempt to explain it.

Pollen data collected from locations around the edges of the Olympic Peninsula indicate that the forests of the peninsula were more open (grassy prairies were larger and probably more abundant) prior to about 5,000 years ago than at present because of a warmer and drier climate (Whitlock 1992). Fossil pollen collected from a bog at Slab Camp high in the Olympic Mountains and a few miles east of the northeastern corner of Olympic National Park indicate that forests there were closed between about 5,000 and 3,000 years ago (Gallison 1991, 1994). The available pollen data thus indicate that alpine-tundra habitats may well have been found in rather small, isolated patches during the hypsithermal, much as Pielou (1991) proposes. Of course, this does not mean that mountain goats (or any of the other species listed in table 7.2) were once present in the Olympics and that they died out during the hypsithermal. It does indicate, however, why they might not have been there historically if they were in fact there earlier: patches of suitable habitat became too small to support them.

"Prehistoric overkill by human hunters may have eliminated one or two isolated populations [of mountain goats]" (Lyman 1988a:20). I did not elaborate further in this paper, but I cited Allan Brooks's (1930) paper on late-nineteenth-century big-game conditions in the Mount Baker area of northwestern Washington. Brooks (1930:65) argued that the big-game population of Mount Baker was exceedingly small as a result of its having been "exterminated by the aboriginal inhabitants long before the advent of the white man." He was responding to J. M. Edson's (1930, 1931) belief that white hunters had depleted big-game populations. Brooks (1930:66) mentioned the mountain goat explicitly. Several elderly Native Americans told him of entire "tribes" climbing into the mountains, placing hunters along escape routes from a mountain basin, and others driving the animals to the hunters; "few escaped a well organized hunt," and these drives "had reduced these animals to the vanishing point on all the nearby ranges" (Brooks 1930:66).

The water-shedding and warmth-retaining properties of mountain-goat wool made it a prized item among Indians living along the northwestern coast of North America (Suttles 1990b). It was such a prized resource that some Indian families traditionally owned mountain-goat hunting territories in British Columbia and southeastern Alaska when the first Euroamericans appeared (Suttles 1990b). Overhunting in these areas would have been unlikely because herds could be replenished by immigrants from herds located farther east, where goats seem to have been hunted less intensively. Overhunting of the more isolated post-Pleistocene (postulated) population of mountain goats in the Olympics would have been much more likely. Brooks (1930:65) apparently was aware of the impact of prehistoric hunters on game populations, for he asked: "How many realize that the Yellowstone Park was a comparatively gameless region [prior to the coming of white men]?" The prehistoric gameless condition of that park has now been well documented (Kay 1994; the herds of elk there that prompted the writing of the Leopold Report in 1963 were a product of twentieth-century game management).

CONCLUSION

The Olympic mountain goat controversy hinges on differing views of the historic biogeography of mountain goats. While I have not shown in this chapter that mountain goats did in fact colonize the Olympics prior to 1925, I find the arguments presented by park biologists that they did not colonize the Olympics to be flawed and not supported by the evidence.

Houston and Schreiner (1993, 1994a) have done the scientific community a service by summarizing many of the available data on endemic organisms in the Olympics. But documenting endemism is only the first step in writing a biogeographic history. They do not consider the fossil evidence, especially for plants, that might provide insights into the biogeographic history of the organisms. Apparently, they believe that the biogeographic isolation of the Olympic Mountains "allowed the evolution of eight species of plants found nowhere else in the world" (Wright 1996:22). But they ignore the fossil record and fail to acknowledge the potential of a genetic test of the mountain goats. They perform none of the typical analyses of insular faunas. Such analyses may be beyond the scope of their summary, but much research and analysis remains to be done prior to deriving a strongly based conclusion that mountain goats were never present in the Olympic Mountains prior to 1925.

Houston and Schreiner's use of lists of mammals missing from various insular locations in the Pacific Northwest is flawed because the lists are inaccurate; therein lies part of the apparent strength of their argument. This is unfortunate because such errors are perpetuated in attempts to justify eradication of the mountain-goat population. It would be a major travesty in "science" to agree to remove goats from the park at this time based on such incorrect information and the usual evidence of the impact of the goat population on the ecosystem. The efforts of Houston and Schreiner, Schultz, and Schalk have taken place half a decade after the park managers made up their minds to reduce the mountain goat population significantly (Olympic National Park 1987a); this suggests that the research is being undertaken in a reactive rather than proactive context: a crisis has been identified. These research efforts are seriously flawed. Scientifically based management decisions demand better.

8

MANAGEMENT OF WHITE GOATS

According to the 1991 Resources Management Plan for Olympic National Park, "the problem of mountain goats was identified as a threat to the integrity of park ecosystems" initially in the park's 1970 Resources Management Plan. On May 3, 1995, Polly Dyer, president of Olympic Park Associates, a "watchdog" group of private citizens "dedicated to preserving the wilderness integrity of the park" (McNulty 1996:218), stated that "Olympic Park Associates and the Klahhane Club ... were the first to observe and bring to the attention of the National Park Service the extreme damage to endemic Olympic plants and fragile alpine soils being caused by the expanded numbers of the non-native mountain goats." Dyer indicated that her group notified the NPS in 1967. Thus, one would think that research on the goats and how they were to be managed would have begun in the early 1970s. It did not begin in any intensive or extensive fashion, however, until 1977 (Stevens 1983). I have reviewed much of that research in previous chapters. The final issues to address concern the management plans that have grown from that research, the alleged effects of the implementation of the preferred plan, and the role of misinformation and advocacy in mustering public support for the preferred plan.

MANAGEMENT PLANS AND ENVIRONMENTAL IMPACT STATEMENT

Houston, Schreiner, Moorhead, and Olson (1991:87; see also Carlquist 1990) indicate that management of the mountain goats in the park began in 1981 with an "experimental management program" (Olympic National Park 1981). The second phase of that management was "initiated in 1988 under provisions of the 1987 environmental assessment" (Houston, Schreiner, Moorhead, and Olson 1991:87; see Olympic National Park

1987a). The January 1992 issue of the *IGMT Newsletter* (no. 1) indicates that a draft environmental impact statement (EIS) was being prepared in which various management alternatives would be described and made available for public comment. Seth Tuler and Chuck Janda (1991:15) state that in late 1990–early 1991 "only two alternatives are being considered: terminating all goat removal operations and allowing the population to return to pre-removal levels, or eliminating the remaining goats by shooting." Janda was chief park ranger at Olympic National Park at the time.

The June 1992 issue of the *IGMT Newsletter* (2:6) indicates that the draft EIS was to be completed during "Summer 1992" and would be published in "Fall 1992." The draft EIS was not published by that date. Late in March 1994 Olympic National Park distributed a news release: "A draft Environmental Impact Statement proposing a program to manage Olympic National Park's population of exotic mountain goats is now expected to be released in late summer or early fall." The Draft EIS (Olympic National Park 1995) was not released until March 1995, however.

The 1987 Management Alternatives

The report "Environmental Assessment: Mountain Goat Management in Olympic National Park" released in 1987 contains descriptions of four "proposed" alternative management plans (Olympic National Park 1987a:50). Three of these alternatives were "intended to slow or stop soil erosion caused by goats and to allow native plants to reestablish the patterns and composition that existed prior to the influence of mountain goats" (Olympic National Park 1987a:50).

Management Alternative 1 is labeled the "no action" alternative (Olympic National Park 1987a, 1987b). Basically, biologists would cease all active management actions. The goats would be allowed to disperse freely throughout the Olympic Peninsula, and their population would be allowed to grow to whatever level it might attain "naturally." Humans would not interfere regardless of the influences of trampling, wallowing, and grazing by goats on the alpine and subalpine ecosystems of the park.

Management Alternative 2 is labeled the "control" alternative (Olympic National Park 1987a, 1987b). Biologists would reduce every subpopulation of mountain goats within the park by 70 to 80%. This would require continued intervention by park biologists over subsequent years to ensure maintenance of the reduced total population at the desired level and would allow some of the areas influenced by heavy goat use to return, theoretically, to more or less goatfree condition.

Management Alternative 3 is the park's "Preferred Management Alternative." Implementation of this alternative would lead to total elimination of mountain goats from "the majority of the park"; in particular, goats would be removed from the "interior core of the park" (Olympic National Park 1987a:52). Goats would remain along the eastern periphery of the park, "where mountain goats in adjacent national forest are managed to maintain a population large enough to sustain an annual sport hunt. The mountainous terrain and widespread presence of goats along the eastern portion of the peninsula offers numerous situations where animals can freely move in and out of the park" (Olympic National Park 1987a:52). As conceived by park biologists, Management Alternative 3 would result in "about 77% of the area presently occupied by goats eventually [being naturally] restored to the type of alpine vegetation patterns that existed before goats moved into the area" (Olympic National Park 1987a:53). Further, it would maintain the hunted population desired by the Washington Department of Wildlife and Olympic National Forest while simultaneously "significantly reversing environmental damage" caused by the mountain goats.

Management Alternative 4, the "Elimination" alternative, would involve total eradication of the mountain goats from the Olympic Peninsula (Olympic National Park 1987a, 1987b).

The "environmental assessment" report (Olympic National Park 1987a) describing these management alternatives was released in 1987 and circulated for public comment. At the same time, a nine-page pamphlet summarizing the management alternatives and answering some often-asked questions about the controversy was printed and distributed (Olympic National Park 1987b). According to Brad Carlquist (1990:15), "Olympic National Park held 16 public meetings over a three-month period in late 1987 and early 1988 to discuss research results," and the public "was given the opportunity to learn more about the proposed management alternatives." The public was provided with a "slide/tape presentation," a "brief management history," an overhead-projector presentation, a nine-page pamphlet (Olympic National Park 1987b), and the eighty-eight-page environmental-assessment document (Carlquist 1990:15). This "open planning process" included consideration of various "social concerns of groups ranging from the Native Plant Society to hunters" and resulted in Olympic National Park choosing their preferred management alternative for implemention (Carlquist 1990:16, 14).

The preferred Management Alternative 3 was first implemented in 1988 and continued in 1989. During the first year of implementation, 80 mountain goats were captured; 7 died for various capture-related reasons and the remaining 73 were transplanted to other areas in the Pacific Northwest. The management alternative called for the removal of 120 goats in 1989, but only 67 were captured; 13 died from capture-related trauma and the remaining 54 were transplanted (all figures are from Houston, Schreiner, Moorhead, and Olson 1991).

The preferred management alternative was originally "scheduled to run through 1992" or until after completion of "a full Environmental Impact Statement" (Houston, Schreiner, Moorhead, and Olson 1991:87, 91). It did not do so, however, because the live-capture operation making up the major part of the management alternative was (in 1990) "canceled by the superintendent of Olympic National Park when two independent assessments indicated that continued efforts posed unacceptable risks for the capture team" (Houston, Schreiner, Moorhead, and Olson 1991: 90–91) and because the live-capture program "just seemed inhumane" as the rate of inadvertent mortality of mountain goats increased (park ranger Paul Crawford, quoted in Luoma 1993:37). The potential that implementation of Management Alternative 3 might result in the accomplishment of Management Alternative 4—complete elimination—was not explicitly recognized by park researchers in the late 1980s but is acknowledged in the Draft EIS (Olympic National Park 1995).

The 1995 Draft EIS

The Draft EIS (Olympic National Park 1995:27) lists three management alternatives that are "based on public and agency issues and concerns." Management Alternative 1 is the preferred one and involves total elimination of mountain goats from park lands "using helicopter-assisted shooting by authorized representatives" (Olympic National Park 1995: 29). Under this alternative, goat carcasses "would be left in place to recycle into the ecosystem" unless located near a trail, road, or other location frequented by humans. The response of plants to goat removal would be monitored. Management Alternative 2 is termed the "no action" alternative. "No goat population management or control action would be taken," and it is supposed that the goat population would "increase to some equilibrium level" (Olympic National Park 1995:31). The net result would be "increased risks to vegetation communities and rare plants caused by increasing goat numbers." Management Alternative 3 would involve elimination of all goats by a combination of live-capture and transplanting the first year and helicopter-assisted shooting in later years.

A great deal of thought went into these alternatives. Numerous variables were considered, such as the time and money necessary to attain their respective goals and the ecological consequences of each. Thirteen other alternatives were considered but rejected as impractical or too costly. Basically, the three retained management alternatives of 1995 are similar to 1987 alternatives 1 (no action) and 3 (preferred total removal from the park), with the twist that the last has two expressions in 1995: removal only by shooting and removal by capture and shooting. Unlike the 1987 plan, the 1995 preferred management alternative acknowledges that "the goat population outside the park may or may not remain viable long-term" (Olympic National Park 1995:50). I suspect that the park would prefer that the population outside the park disappear as well, because then they would not have to patrol the border of the park to keep intruding goats out. From a cost-benefit perspective, that is reasonable.

I do not believe that the research that park scientists have accomplished allows us to make a well-informed choice of alternatives for two reasons. First, NPS's published definition of native species means that if goats *ever* occurred naturally there, the extant goats must be considered native. However, immediately post-Columbian biotas are the goal (Houston and Schreiner 1995; Wagner et al. 1995), so we must determine if goats were present or not between 1492 and 1925. Thus far, we can only make this determination for the 1825–1925 period, in an interpretation founded on contradictory historical evidence and limited ethnographic evidence.

Second, most of the research has not been independently peer reviewed. The Draft EIS (Olympic National Park 1995:12, 23) indicates that "40 goat-related projects ... have resulted in approximately 80 reports ... during the past 30 years.... Many of these reports have been published in scientific journals." Research did not begin in earnest until about twenty years ago. I screened the data in the appendix to ensure that each report was tallied only once (instead of both a published and unpublished version), to omit all reports authored by nonpark employees, and to include only those titles with direct relevance to the controversy that contain scientific data. Fifteen reports were published in scientific journals, sixteen reports were published in popular outlets but do not contain scientific data (e.g., *IGMT Newsletter* issues and articles in *Park Science*), and there are fifty-seven unpublished reports. Many of the pertinent data on sterilization and contraception (Warren et al. 1992), on archaeology (Schalk 1993), and on wallows are not published. The lack of published research is apparently not unusual throughout the NPS (Wagner et al. 1995:97–98).

On March 8, 1996, Olympic National Park issued a news release in which Park Superintendent David Morris indicated that completion and release of the Final EIS, originally to have been released in late 1995, would be delayed until fiscal year 1997. In this news release the delay was not attributed to budget constraints, although it later was (see below). Morris did not mention that a clear majority of the public comments on the Draft EIS received by the park opposed the plan to eradicate the goats (Wright 1996). Nor did he mention in this news release that the Olympic milkvetch, one of the endemic and supposedly rare plants threatened with extinction by the goats, was no longer considered to be in jeopardy. The Final EIS had not been released as of October 1997, in part because its planned release was halted by actions that took place in December 1996 (see below).

ECOLOGICAL RESTORATION

Olympic National Park (1987b:3, 6) stated that "the Olympic subalpine ecosystem can be restored to its original condition if goats are removed" and that "recovery of some areas would require active rehabilitation." Five years later botanist Lawrence Bliss was quoted in the second issue of the *IGMT Newsletter* (2:5) as stating that "plant communities have changed dramatically following experimental removal of goats from [impacted] area[s]." But we do not know if those changes produced the pregoat (pre-1925) condition because we do not know what the pregoat condition was. The potential for recovery of pre-1925 conditions thus is unclear. In some unspecified cases human intervention ("active rehabilitation") will be unnecessary; in others such intervention may be mandatory to recreate pregoat conditions, according to park biologists. Some studies (e.g., Curtin 1995) suggest that human intervention may be necessary in every case involving alpine and other montane habitats.

What will human intervention involve and when will it be employed? To borrow the title from Alston Chase's (1987) book, intervention by park biologists would truly seem to be "playing God" in the Olympics. How can anyone suggest that removal of the goats will result in pregoat conditions when, as most people admit (but not loudly), the pregoat conditions of the park are unknown? One could, of course, also argue that to recreate a "natural" ecosystem dating to, say, the sixteenth century would require the park to allow Native Americans to resume traditional hunting, fishing, and gathering activities in the park.

The alpine ecosystem of Olympic National Park has included mountain goats for at least seventy years. It is extremely doubtful that removal of the remaining goats will cause the ecosystem to revert to some (unknown) pregoat condition. Have genetic changes in plants resulted from the presence of mountain goats? Might some new form of plant be on the horizon as the goats continue to graze the alpine meadows? Is not one goal of the maintenance of natural areas the maintenance of biological diversity? What geological processes have been irreversibly altered? Natural areas are also supposed to maintain a diversity of natural processes (Smith et al. 1993; Wagner et al. 1995). These questions highlight the long-range planning process that must be considered if the preferred management alternative in the 1995 Draft EIS is to be implemented. Have the park managers wrestled with these questions or have they simply determined that eradication of the mountain goat population is the (short-term) goal? It is clear that their strategy of resource management is not as rigorous or as robust as it might be (Peterson et al. 1994). Is this perhaps part of the problem?

NPS policies dictate how to deal with "exotic" organisms, and establishing that an organism is in fact exotic is the only step requisite to removing it from park lands. To find that the mountain goats are exotic and that they are having a significant negative impact on the alpine and subalpine ecosystem simply because that is what one wants to find violates every tenet of scientific procedure (e.g., Chamberlain 1897). As Wagner et al. (1995:113) note, "Resource-management agencies [such as NPS] are constituted to advocate the social values that they are chartered to satisfy. In this day and age they also employ science to enlighten their policy decisions. If science is to serve management efficiently, it must strive for objectivity, which requires distancing itself from management's value positions." Although science is supposed to be objective, this does not mean, of course, that park scientists cannot have opinions or advocate particular viewpoints that are founded in science.

ADVOCACY

The role of advocacy in scientific research performed in the service of natural resource management has become increasingly important in recent years. Two collections of papers (Noss 1996; Wagner 1996) and a major article on the topic (Woodward and Goodstein 1996) published by Sigma Xi (the Scientific Research Society) appeared in 1996. In general, all of these papers underscore that scientists must clearly distinguish between advocacy and science (see also Lee 1993; O'Brien 1993; Sweeney and Stangel 1995). Science can and should indicate what is to be advocated; what one wishes to advocate should not, however, influence the scientific research, evidence, and results.

The role of science in modern wildlife management is clear: it is a tool used to ensure that a workable protocol that will produce a desired result is chosen for implementation. Thus, the first issue addressed is the choice of the kind of result one desires; in the case of natural resources, one must choose how those resources are to be used. Do people want, say, a wildlife preserve that contains only native species or one that contains an abundance of large mammals that may be harvested at a sustainable rate? Answering such questions strongly influences wildlife management policies and demands that one make a choice. Science can play several roles. It can help decide which species are native or what an acceptable harvest rate is. But it can also be called upon to help make the choice.

Scientific advocacy, for example, plays an important role in determining what constitutes good ecological restoration (e.g., Higgs 1997). Natural resources may be put to many uses and thus have many values (e.g., Gray 1993; Higgs 1997; Lemons 1987; Wagner et al. 1995). Particular values tend to reside within different subsets of the human population, with each subset desiring to use the resources differently. Here science may begin to shift from being a tool that generates information that allows us to weigh the strengths and weaknesses of alternative choices to being a tool that can be used to advocate a particular choice.

Because scientific research can be used as a tool of advocacy, it can be misused. It is well documented that—unlike the conventional (Brush 1974) or perhaps Hollywood image of an individual in a white lab coat searching for truth—scientists in general often tend to resist new theories or new data, particularly when they jeopardize pet ideas (e.g., Barber 1961). It is, of course, the nature of scientific investigation and a healthy phenomenon that multiple investigators will disagree over the significance of some newly discovered data or newly proposed theory. Disagreement ensures that fewer inaccuracies will be incorporated in our general knowledge than if no argument takes place. Disagreement signifies that we are struggling to figure out precisely what it is that we think we know or why we think we know it.

Things get sticky, however, when advocacy enters the arena. Disagreement over data or theories can be attributed to a scientist's being wedded to the advocated position and thus blind to the significance of contradictory information. Admitting that one might have been incorrect in this advocacy is especially dangerous when one's job security depends on being correct. Part of the problem also resides in what has been termed the historical inertia of an idea. Once a theory or hypothesis has become a part of the accepted wisdom of a science, it is difficult to get rid of it. The historical inertia of the idea that mountain goats never have been native to the Olympic Peninsula seems to be part of the driving force behind the position advocated by Olympic National Park biologists. But the other lesson of the history of science is that a "scientist must be brave enough to question and criticize anything his teachers or his society may tell him, at the risk of ostracism, denial of financial support or worse.... There is only one established dogma in science—that scientists do not blindly accept established dogma" (Brush 1974:1165).

WILDLIFE MANAGEMENT, SCIENCE, MISINFORMATION, AND THE PUBLIC

The final EIS was originally scheduled for release during the winter of 1995–96. It was not released at that time, however, because of the shutdown of the U.S. government early in 1996 and the resultant lack of a budget for the National Park Service, according to Park Superintendent Morris's statements in the park newsletter for December 1996. U.S. Representative Norm Dicks (D), whose district lies in western Washington state, met with personnel representing the Fund for Animals and Olympic National Park personnel in December 1996. He listened to both sides of the controversy and concluded that the park was acting too quickly in seeking to remove (in particular, to shoot) the goats. Western Washington newspapers reported the meetings, and on January 1, 1997, Park Superintendent Morris announced that the park was attempting to "come up with a compromise that satisfies [Representative Dicks's] concerns." Olympic National Park's newsletter (accessible on the World

Wide Web at http://www.nps.gov/olym/) for December 1996 (1:2) contains the following statement by Morris: "At the urging of Congressman Norm Dicks and in consideration of public concern with this issue, the National Park Service is continuing internal discussions regarding mountain goat management. Park staff continues to work towards a solution which recognizes the widespread public interest, as well the Congressman's concerns." It seems, then, that advocacy of a particular point of view by an individual who is sufficiently powerful politically can alter what is said to be a scientifically founded management objective. But advocacy in this case does not stop there.

From the perspective of the scientists at Olympic National Park, there should not be a controversy at all. Park resource-management specialist Paul Crawford (1993:15) stated that the damage to the alpine ecosystem created by the goats is "serious" and that the research demonstrating this not only has been "thorough" but has also been peer reviewed; it has "undergone thorough academic scrutiny" and "technical evaluation by subject matter experts." The peer reviews so boldly underscored by Crawford have not been infallible, nor have many of them been independent (see below). Many errors of omission and commission appear in the documents produced by park personnel. Perhaps because most of these documents are "NOT FOR PUBLICATION" and thus have not been subject to peer review, those errors have been perpetuated in popular articles in magazines that are widely circulated to the public. This is a rather subtle but unacceptable version of advocacy.

Articles in *National Geographic* ("Alien to Olympic Park, Mountain Goats May Go" 1992), *Sports Illustrated* (Gutterson 1993), *Sunset* ("Between a Rock and a Hard Place" 1995), and *Scientific American* (Wright 1996) by individuals who are not park employees are too short to review the relevant data and arguments. Other articles, such as those appearing in *Animals* (Sleeper 1990), *Wildlife Conservation* (Luoma 1993), and *National Parks* (Weingrod 1994), while including interpretations of the available data, are too superficial to allow more than emotional responses (e.g., Bishop and Martin 1994). Emotions are important in such issues (Scheffer 1993a), but they seem to be irrelevant in the eyes of the NPS. The authors of most of these articles read some of the park literature and interviewed park personnel. They did not always interview individuals aligned on the other side of the controversy. This, too, is a rather subtle but unacceptable form of advocacy.

The delay of the release of the final EIS through the 1996 intervention of Congressman Dicks resulted in a flurry of advocacy. On January 8, 1997, the following statements (among others) appeared in the section on "Action Alerts" maintained by Olympic Park Associates on the World Wide Web (http://www.halcyon.com/rdpayne/opa.html): "Letters are needed to Congressman Norm Dicks of Washington protesting his politicization of a scientific problem....The Park Service has clearly documented significant and measurable impacts to Olympic's alpine and subalpine ecosystems, directly as a result of the presence of introduced mountain goats....it is incumbent upon our elected officials to allow Park Service officials, with the aide of scientific research and public *participation*, not *coercion*, to preserve and protect the ecological integrity of our National Park System." Is Dicks's involvement "politicization of a scientific problem" because he is a politician?

The Olympic Park Associates World Wide Web page also includes a statement of their official position on the controversy (authored by president Polly Dyer) and synopses of Schreiner's, Schultz's, and Schalk's research, similar statements by the Cascade Chapter of the Sierra Club, and the Olympic Park Associates newsletter, *Voice of the Wild Olympics*. The April 1997 issue (5:1) of the newsletter proclaims that to argue goats are native is "bull" and that Congressman Dicks's efforts "at the eleventh hour" represent a "torpedo[ing] of more than fifteen years of exhaustive research." Olympic Park Associates' advocacy for the park's position that the goats should be exterminated is clear.

Advocacy for the park's view takes other forms as well. The third edition of a free booklet published in 1994 by the American Park Network and available in various locations around the park lists eleven species as missing from the Olympics (coyote not included; all others in table 7.2 are listed). This booklet states that "oddly [these eleven species are] not in the Olympics. Scientists think that they probably could not survive the Ice Age in the Olympics, or else they never reached the geographically isolated region" (*Mount Rainier and Olympic National Parks Magazine* 1994:82). Park visitor center bookshops sell Tim McNulty's (1996:58) book *A Natural History Guide: Olympic National Park*, which states that goats are nonnative and exotic, that extant goats are "changing the composition of the park's alpine and subalpine plant communities," which had "evolved in the absence of [mountain goats]," and that the goats pose "a threat to the ecological integrity of the park." Similarly, park visitor center bookshops have copies of Henry C. Warren's (1982) book *Olympic: The Story behind the Scenery* for sale. This book lists the eleven mammals missing from the park (table 7.2), which are said to be nonnative, and says that the mountain goats "represent a serious management problem because they harm the ecology" (Warren 1982:34). This form of advocacy, too, is unacceptable. But it has another, more dangerous manifestation as well: censorship.

A "coffee table book on the Olympic Peninsula" published by Mike Phillips was "banned from Olympic National Park" and rejected by visitor center bookshops because it contains the statement, "Despite eyewitness accounts by two of the three original exploring parties who visited the Olympic Mountains, Olympic National Park officials consider mountain goats to be exotic animals" (Phillips 1996). Phillips goes on to state that when the methods of park biologists "have been challenged, they have thrown up the barricades. When contradictory evidence has been brought forward by citizens, they have turned a little switch that prevents bothersome sound waves from reaching their ears. This is how a park bureaucracy banishes citizens. It tells them they can be tourists, but they can't be partners in managing a park that belongs to all of us." Sour grapes? Perhaps, but certainly warranted. This is an example of how scientific research can be used, or not used, by park scientists to advocate a particular viewpoint.

Although Olympic Park Associates are quick to reference the scientific research performed by park researchers, there is no evidence indicating that they have evaluated the quality of that research. Their lack of understanding of science is also apparent in their advocacy of the removal of the Elwha River dams. The presence of one of those two dams within the boundaries of the park is not natural (one dam is outside park boundaries). But for the associates and park researchers to believe that by removing the dams we will "fully reverse an environmental mistake" (Wunderlich et al. 1994:18) and restore a natural, predam ecosystem is nonsense. As the fisheries biologists indicate, "six of the 10 native Elwha anadromous fish stocks are either extinct or acutely reduced" and thus "nonnative introductions may be necessary for some stocks" (Wunderlich et al. 1994:14, 16). A perfect reversal of ecological processes cannot result from the use of different gene pools, although an approximation might. But this ignores the irreversible aspects of the dams' existence for over more than fifty years.

After the goats and the dams are removed (if that is the way the cards fall), will park biologists turn their attention to removing other organisms from the park that seem to be exotic, such as red foxes, coyotes, and various birds? The introduction of wolves is being considered (McNulty 1996:223–24), but I wonder which gene pool they will use. These are separate issues, yet they, too, raise the question of what the long-term park management plan entails.

The result of the Olympic National Park mountain goat controversy thus far has been some rather bold assertions in the published literature. The goats are exotic because there is no zooarchaeological evidence that they were native, when the prehistoric record is insufficiently known to warrant any statement regarding their prehistoric presence or absence. The survival of various rare endemic plant species is threatened by the goats, when the goats can only be accused of killing less than one-half of one percent of the plant population and insects seem to be an equally serious plant predator. The goats are solely responsible for the initiation and exacerbation of erosion of the fragile alpine ecosystem, despite the presence of other large and small mammals—including humans—and the late-nineteenth-century presence of domestic livestock. The white lies go on and on.

THE FINAL RESOLUTION?

On June 16, 1997, Olympic National Park issued a news release that stated, in part:

> The Secretary of the Interior, in consultation with the Secretary's science advisor and appropriate professional societies, will appoint a number of recognized experts to review existing scientific information about mountain goats in Olympic National Park.... Up to 12 independent scientists, who are yet to be named by the Secretary, will examine all mountain goat research done to date, including history of goats on the Olympic Peninsula and the effects of goats on soils, vegetation, and natural processes. They are expected to present their findings to the Secretary of the Interior within 18 months. The public will have an opportunity to review and comment on these findings.

The manner in which an earlier supposedly independent peer review was orchestrated by Olympic National Park raises doubts as to whether the goats will get a fair hearing, the evidence will be thoroughly scrutinized and independently evaluated, and the final decision will be truly science-based.

The *IGMT Newsletter* (June 1992) states that, "since questions have been raised, over the next few months the IGMT will contact reputable independent scholars to evaluate the historical, archeological, anthropological, and paleontological records." This review "was partly in response to an article by Dr. R. Lee Lyman [1988a]" and involved nine experts "selected" (Olympic National Park 1995:225), at the request of the park, by the Forest History Society, the American Society of Mammalogists, and the History of Science Society. The park also asked the archaeologist in charge of the NPS's own Archeological Assistance Division to select three archaeologists to participate in the review process. Each of the nine experts had three weeks to review four documents and to prepare a report evaluating those documents. They were paid $500 for their efforts; all of the requisitions were signed by Olympic National Park archaeologist Paul Gleeson. The four documents reviewed were Schultz's (1993) review of the historical evidence for pre-1925 mountain goats on the Olympic Peninsula (73 pages); Schalk's (1993) review of the ethnographic and archaeological evidence for native mountain goats (59 pages); the Houston and Schreiner (1993) manuscript on the biogeography of the Olympic Peninsula (27 pages); and my paper (Lyman 1988a) outlining a conjectural biogeographic history (11 pages; summarized in chapter 7). The dates for the cover letters soliciting the reviews from the experts range from June 9, 1993, to July 27, 1993. In my view, this evaluation should have taken place prior to the mid-1980s when the program to remove the "exotic" goats was initiated.

Material sent with the four documents to the reviewers says, "We ask you to review the enclosed material." In a letter Park Superintendent Finnerty states that the Houston and Schreiner (1993) document "was sent to the [nine] reviewers for background information and was not a primary part of the material for review" (Finnerty letter to Lyman, March 10, 1994) and that "the Houston and Schreiner document was provided in draft form as background material for the other review items" (Finnerty letter to Lyman, May 13, 1994). But the reviewers were specifically directed to "review" all four of the documents. When I asked Finnerty about this apparent discrepancy, Olympic National Park personnel stopped discussing issues related to the controversy with me. During this period I also offered to document the errors I perceived in various park documents (outlined in chapters 6 and 7). They have never requested that I do so. These facts make me wonder if the public will be allowed to comment on the 1997–98 review process.

The first two paragraphs of the background material sent to the nine reviewers in 1993 read as follows:

> Early mammalogists considered that the mountain goat (*Oreamnos americanus*) was absent historically from the fauna of the Olympic Mountains, as were several other species of mammals present in the nearby Cascades. Goats are known to have been introduced to the Olympic mountain range in the 1920's. The National Park Service has viewed goats as an introduced species in Olympic National Park. The interpretation that goats were absent historically has been challenged, and this prompted our review of the original premise.
>
> We ask you to review the enclosed material and consider the question: How strong is the evidence that goats were absent historically? We are well aware that the historical sciences cannot prove the negative, and that your assessment will necessarily involve weighing the relative strength of the negative evidence. We are also interested in your assessment of the likelihood that goats were absent during the late Quaternary, based on the available evidence.

The citation states four times and implies twice that goats were absent from the Olympics prior to 1925. The possibility that goats were present prior to 1925 is not mentioned. This hardly makes for an independent review; the hope of the park, which is paying for the review, is explicit. The first sentence quoted above indicates that "early mammalogists" believed goats were absent. This is a biased appeal to authority because it does not indicate that only *some* biologists believed goats were absent; two reputable biologists, Fannin and Grinnell, suggested that goats were present. The first sentence also does not indicate that the "early mammalogists" produced demonstrably incomplete lists of the mammals present in the Olympics and made various biological errors. I wonder if a less slanted introduction will preface the materials reviewed in 1997–98 (see "A Final Note" below).

The reports by Schultz, Schalk, and Houston and Schreiner examined by the reviewers are secondary literature. They provide summaries of the primary literature written from a particular perspective that could be different from that of the original authors. Significant data might not be mentioned, and various information might be cited out of its appropriate context. From my reading of much of the primary literature reviewed by Schultz, Schalk, and Houston and Schreiner, it is clear that the nine reviewers could not have completely accurate or objective knowledge of what the original authors said or thought. None of the members of the review team, as far as I know, had been involved with the controversy until asked to participate in June 1993, so I doubt that they knew the literature in sufficient detail to offer completely thorough reviews of the primary evidence.

The reviewers in 1993 by and large made evenhanded judgments *of the evidence made available to them.* It is the incomplete and secondhand nature of this evidence that worries me. The very limited time frame they were given to review that evidence virtually precluded doing any research on their own, though several of them did refer to relevant research they were already familiar with. Hence the nine reviewers did not consider the documents they examined in sufficient detail. For example, none of them detected the errors in Houston and Schreiner (1993; see chapter 7). Further, none of them mentioned the third primary reference to the pre-1925 presence of mountain goats in the Olympics (Fannin and Grinnell 1890c). None of them noticed that C. Hart Merriam was apparently using secondhand information when he stated that it was "said" that mountain goats were not present in the area (as quoted in Schultz 1993). And none of them noticed that Schalk (1993) overlooked several archaeological sites or that Schultz's (1993) analysis was biased.

Several of the reviewers found the reports of Schultz and Schalk to be "thorough," but these reports are not thorough (see chapter 6). Several reviewers found the archaeological evidence convincing, but they did not evaluate that evidence in any manner, nor did they have all of the available evidence at their disposal. At least one reviewer incorrectly believed that the zooarchaeological record for the Olympic Peninsula included "reports of *all* mammals known from the peninsula" (emphasis added). That is inaccurate because we have not yet found any remains of the Olympic marmot, a point emphasized by Schalk (1993); nor have we found remains of several other species known to occur on the peninsula.[1] Perhaps most indicative of the lack of primary knowledge on the part of

[1] I am presently (summer of 1997) identifying the animal remains recovered from a 2,000- to 4,000-year-old archaeological site in the northeastern portion of the Olympic Peninsula. I have identified three genera of mammals not previously recorded in archaeological collections from peninsula sites: *Scapanus*, *Microtus*, and *Peromyscus*. These genera are not indicated as having been so identified in table 7.6, and it is not clear at this time if the remains of these animals are several thousand years old or are intrusive to the site sediments and thus are relatively recent in age.

the nine reviewers is the fact that one of them still believed in 1993 that 1,200 goats were then running around the Olympic Peninsula. But several reviewers (including two of the archaeologists) agreed with Schalk (and me) that the zooarchaeological record was probably not representative of the late Quaternary alpine fauna of the Olympic Peninsula or at least agreed that this record as described by Schalk did not seem to provide a firm basis for concluding that mountain goats were never present in the Olympics.

The histories of the two instances of "independent" review are thus far quite similar: questions are raised, professional societies are asked to help select reviewers. I am hopeful that the histories will not be similar in their later phases. If they are, then we can expect the reviewers to be told at the start what they should conclude or at least what the park thinks they should believe. There will be no review of "*existing* scientific evidence"; that evidence will be screened, and only selected portions of it will be provided to reviewers in the form of secondhand documents. The reviewers will be handpicked, some at the suggestion of individuals who owe their livelihood to the NPS. The reviewers will be given a limited amount of time to devote to researching critical issues. And the public's comments, or at least those that disagree with the park, will be ignored.

A FINAL NOTE

The preceding section was written in the early summer of 1997. At about that time, the NPS drafted a "Request for Quotations" seeking bids by scientists to perform the new (1997–98) review of the scientific evidence concerning management of mountain goats in Olympic National Park. It is my understanding that the NPS, rather than Olympic park personnel, are setting up the review procedure in an attempt to displace the kinds of doubts about objectivity that were expressed regarding the 1993 review. The draft statement requesting proposals and bids to undertake the new review, so far as I know, had not been released as of November 15, 1997. Should that statement be released, I fear the outcome is in some respects fore-ordained.

The draft statement characterizes the new review as an "independent scientific review" and that the review will be of the "de novo" sort. The draft statement explicitly states that "Reviewers must be impartial regarding the ultimate decision among various mountain goat management options, and must not have expressed a position on management of mountain goats at Olympic National Park which would indicate an inability to be impartial." I find this a good criterion, but believe that its purpose is circumvented by other wording in the draft statement. First, the draft statement indicates that the reviewers "will be provided bibliographies of scientific material known to be relevant to this issue" (all quotations in this section are from the draft "Request for Quotations"). Those bibliographies will be taken from four documents produced by Olympic park personnel: Houston, Schreiner, and Moorhead (1994), Olympic National Park (1995), Schalk (1993), and "additional references included in the draft Final Environmental Impact Statement for Mountain Goat Management, Olympic National Park, August, 1977." In other words, the reviewers will be provided with only those references that park personnel thought useful. Although I have not examined the 1977 "draft Final Environmental Impact Statement," it is clear that the other bibliographies do not contain references to all relevant documents and thus the reviewers will have to do some digging to find, for example, references cited in *White Goats, White Lies.* The draft statement also indicates that the reviewers are expected to "locate any other relevant scientific material and will solicit such material from professional sources and the general public" as the goal of the review is to examine "all relevant scientific material." I can only hope that the reviewers will be given names, addresses, and phone numbers of individuals with view points different from those of Olympic National Park personnel.

The second reason that reviewer impartiality may be circumvented is that the reviewers are explicitly required to "conduct a site visit to Olympic National Park and the Olympic Field Station of the Forest and Rangeland Ecosystem Science Center, USGS–Biological Resource Division, for staff briefings regarding field research conducted on this issue, and/or first hand assessment of field conditions to complement research review tasks." This is reasonable; the reviewers will talk to those individuals with first-hand knowledge of the research that park personnel have undertaken. Nowhere in the draft "Request for Quotations" is it even hinted that the reviewers have the option of visiting other factions with other view points based on other data. I believe they should be encouraged to do so to ensure that all issues and data involved have a thorough hearing.

Of course how the new independent review will actually take place is unknown, as are its results. If the draft "Request for Quotations" document is an indication, the scepticism I felt early in 1997 regarding the new review seems well founded. More white lies? Hardly. But certainly the structure of the new review seems to have been carefully crafted. My questions about procedures for managing publically owned resources remain.

CONCLUSION

I entered the Olympic National Park mountain-goat controversy, before it really was a controversy, because I saw an interesting biogeographic problem. I stuck with it because I was learning some interesting things about ecology, alpine environments, and mountain goats. I also thought I had found a classic case of Bailey's (1982) "muddling through." I was and still am willing to live with the fact that "we will never have perfect knowledge, and to delay management decisions until we do is a path to inaction" (Wagner et al. 1995:175). Hutchins (1995:1326) refers to waiting for perfect knowledge as the "paralysis of analysis," and I agree. A decision must be made. But what I found in the Olympic mountain-goat controversy was that the scientific data were too often incomplete, contradictory, or inconclusive to support the conclusions of the park biologists. The biologists occasionally admit that their data are not what they should be; yet the public is told that the scientific evidence is beyond question, and park managers and researchers persist in insisting that the goats must be eradicated.

Park researchers often say that questions from outside the park are founded on poor logic, contradict or selectively use available evidence, or are emotional pleas and thus beneath serious consideration (e.g., Houston and Schreiner 1995). I cannot help but wonder why they have largely ignored questions from outside. Is it because "esteem in the NPS is developed by being 'good soldiers' and achieving task goals that support the belief that 'rangers can do anything'" and work with the Olympic mountain goats "is regarded by peers as 'real' ranger activity—the epitome of ranger professionalism" (Tuler et al. 1992:337)?

Olympic National Park, created in 1938 and dedicated in 1946, was recognized internationally on October 26, 1976 (Franklin 1977), when it was declared a Biosphere Reserve under UNESCO's Man and the Biosphere program, "a system of natural areas that preserves the major biomes of the world for research, education, and protection of genetic diversity" (Burger 1987:28).[2] The park was designated a World Heritage Site on October 27, 1981, and the document recommending the park for such designation read in part:

> It is apparent ... that the Olympic National Park is the best natural area in the entire Pacific Northwest, with a spectacular coastline, scenic lakes, majestic mountains and glaciers, and magnificent temperate rainforest; these are outstanding examples of on-going evolution and superlative natural phenomena. It is unmatched in the world.... The main danger to the integrity of the site is, oddly, one of its attractions—the mountain goat. Due to the isolation of the site, mountain goats never dispersed naturally to the Olympics, so their introduction in 1925–29 may be causing significant changes in the natural ecosystem. (quoted in Olympic National Park 1987a:5)

It would be interesting to know who wrote this recommendation, which reiterates the perceived biogeographic history of the Olympic Peninsula as that history was first conjectured by Merriam a century earlier and reiterated by Scheffer (1946). For example, in this recommendation we read that mountain goats "*never* dispersed naturally to the Olympics" (emphasis added). The truth of this statement has not been demonstrated, the evidence mustered in support of it is weak, and the evidence (fossil or zooarchaeological) that might show it to be false has not been collected, despite the fact that park researchers know what that evidence must be.

Systematic research on the goat population only began in 1977 (Stevens 1983), and the production of reports on that research began at approximately the same time. I find the close temporal correspondence of the initiation of research, report production, and the designation of the park as a Biosphere Reserve intriguing. The managers of the park are muddling along, doing their best to maintain or recreate a pristine, natural, pre-Euroamerican ecosystem (whatever that might be). Suddenly, and almost simultaneously, they gain international recognition and begin a very active, very high-profile management program. Could the former be the source of the latter?

---

[2] The date of declaration is incorrectly reported as 1978 in Olympic National Park (1981:7) and in Burger (1987:28). The date is reported as the "mid-1970s" in Olympic National Park (1987a:5).

Certainly, the line in the recommendation stating that the goats "*may* be causing significant changes" (emphasis added) is not said loudly in public. What is being said loudly and often is that the goats *are* having serious and deleterious influences on the ecosystem.

While I do not know that my interpretation is correct, I find the possibility that a political agenda (removing a blemish from an international award) is driving a wildlife-management program disconcerting. Scientific research is perhaps being (mis)used to fulfill and justify that agenda. As I have shown throughout this volume, the quality of the science is not what it should be.

The problem is exacerbated by the manner in which the research results are being used to advocate a position and persuade the public that what the park managers desire is scientifically warranted. At the present time that is difficult to demonstrate. I therefore admonish park researchers to knuckle down to the task of strengthening and clarifying the scientific evidence and to circulate all unpublished reports and documents bearing on the controversy—especially those marked "NOT FOR PUBLICATION"—so that they may be evaluated by all interested members of the public and scientific communities. Those materials must undergo independent peer review if they are to be considered scientific evidence.

As noted in chapter 1, I did not write this book because I believe that mountain goats were present in the Olympics sometime prior to 1925, because my ideas regarding biogeography are being ignored by park biologists, or because I believe that the extant goats should be left alone. Nor am I building an argument that the goats should be left alone if remains of mountain goats older than 1,000 years are found in the Olympics. An argument like this has been presented for various mammal species elsewhere by Quaternary paleobiologist Paul Martin (e.g., 1970). While Olympic National Park biologists have thought that my argument (e.g., Lyman 1988a) paralleled Martin's (e.g., Moorhead 1988b), I do not agree with Martin that the goal of wildlife or park management should be the re-creation of late Quaternary environments, habitats, or faunal and floral communities.

I wrote this book because it seemed to me that the science said to be the sole reason for the decision that the goats must be eradicated was faulty. Some data were not being publicized, and thus the Olympic goats, who cannot speak for themselves, were being misrepresented. Contradictory data were being ignored. Frequently cited pertinent data seemed not to exist. Some data simply did not indicate to me what they were said by park researchers to indicate. Finally, it seemed that the public was being told a series of white lies about the mountain goats in the popular literature.

As I completed the second draft of this volume, I was pleased to read Houston and Schreiner's (1995:208) suggestion that "the National Park Service should consider the advantages of narrower policy guidelines for defining alien and native species; a policy without defined temporal and spatial bounds might lead to a 'whatever feels right' approach to the management of particular species.... We recognize that framing workable guidelines will not be easy and that the attention of ecologists, paleontologists, and conservation biologists will be required." This is precisely the point I made, if less eloquently, in my original paper on the Olympic mountain-goat controversy a decade ago (Lyman 1988a). As I noted then, given NPS definitions of native and exotic species, thousand-year-old mountain-goat remains in Olympic National Park should prompt NPS to "rethink policy issues.... [I]nstances of wildlife management policies being solely based on historic data may have to be re-evaluated once prehistoric data are brought to bear" (Lyman 1988a:22). I still think this is true.

APPENDIX

References listed below were used to construct the graphs in figures 5.1 and 5.2. Some of these were not inspected due to their unpublished status. If a report was cited in two or more other reports, articles, or monographs, it is included here. For those titles that I did not personally inspect, variations in a title cited by others are given in brackets. Where known, the number of pages is indicated. The capital letter in brackets at the end of some references indicates whether a title is tallied as an unpublished scientific report [U], a published scientific report [P], or a published news release that contains minimal scientific evidence relevant to the controversy [P*]. Reports that are tallied as partially published are indicated by fractions [U 1/2, P 1/2].

ABBREVIATIONS

NPS	National Park Service
ONP	Olympic National Park
OSU	Oregon State University
UW	University of Washington

REFERENCES ON MOUNTAIN GOATS

Besser, T. E., and W. G. McGaffey. 1981. An Experimental Sterilization Program for Mountain Goat Population Control. Unpublished report on file with ONP. 16 pp. [U]

Brent, H. 1960. Helicopter Goat Counts. Unpublished memo, Washington Department of Game. 1 p. [U]

Burger, J. 1987. A Mountainous Appetite. *National Parks* 61(1–2):28–31. [P*]

——. 1991. Klahhane Ridge Comparison Photographs. Unpublished report on file with ONP. [U 1/2, P 1/2]

Carlquist, B. 1989. Introduced Mountain Goats in Olympic National Park: Problems and Solutions. UW M.F.R. Professional Paper. 74 pp.

——. 1990. An Effective Management Plan for the Exotic Mountain Goats in Olympic National Park. *Natural Areas Journal* 10:12–18. [P*]

Crawford, P. 1993. Olympic Mountain Goat Update. *Park Science* 13(3):15. [P*]

Driver, C. H., L. L. Bliss, D. K. Pike, and W. A. Pfitsch. 1979. Terrestrial Baseline Surveys, Non-native Mountain Goats of Olympic National Park. [Mountain Goat Habitat Studies] UW College of Forest Resources, unpublished report on file with ONP. 26 pp. [U]

——. 1980. Terrestrial Baseline Surveys, Non-native Mountain Goats of Olympic National Park. UW College of Forest Resources, unpublished report on file with ONP. [U]

Driver, C. H., V. Stevens, and I. Olmstead. 1977. Terrestrial Baseline Surveys, Non-native Mountain Goats of Olympic National Park. [Mountain Goat Population and Habitat Studies] UW College of Forest Resources, unpublished report on file with ONP. 22 pp. [U]

Driver, C. H., V. Stevens, and D. Pike. 1978. Terrestrial Baseline Surveys, Non-native Mountain Goats of Olympic National Park. [Olympic Mountain Goat—Annual Report] [Mountain Goat Population and Habitat Studies in Olympic National Park] UW College of Forest Resources, unpublished report on file with ONP. 154 pp. [U]

Driver, C. H., R. Taber, and V. Stevens. 1979. Mountain Goat Population Studies, Annual Report. UW College of Forest Resources, unpublished report on file with ONP. [U]

Fielder, P. C., and B. G. Keesee. 1988. Results of a Mountain Goat Transplant along Lake Chelan, Washington. *Northwest Science* 62: 218–22. [P*]

Foster, J. W. 1985. Report on Observations of Olympic National Park Goat Capture Operations, 22 June 1985. Unpublished report on file with ONP. [U]

Foster, J. W., B. B. Moorhead, and T. A. Reichard. 1983. Medical Considerations in the Translocation of Mountain Goats. *Proceedings of the Annual Meeting of the Association of Zoo Veterinarians*, pp. 111–116. [P*]

Henderson, J. A. 1984. Goat Salting Impact Study. Olympic National Forest unpublished report on file, Olympia. 26 pp. [U]

Hoffman, R. A. 1987. A Comparison of Population Control Techniques for Mountain Goats in Olympic National Park, Using Field Sterilization Procedures and Computer Modelling. M.S. thesis, University of Idaho, Moscow. 82 pp.

Hoffman, R. A., and R. G. Wright. 1987. A Comparison of Population Control Techniques for Mountain Goats in Olympic National Park, Using Field Sterilization Procedures and Computer Modelling. University of Idaho, Cooperative Park Studies Unit, unpublished report on file with ONP. 81 pp. [NOTE: this and the preceding title are tallied as one in figures 5.1 and 5.2]

———. 1990. Fertility Control in a Non-native Population of Mountain Goats. *Northwest Science* 64:1–6. [P]

Houston, D. B. 1981. A Proposed Experimental Management Plan for Mountain Goats. Unpublished report on file with ONP. 4 pp. [U]

Houston, D. B., B. B. Moorhead, and R. W. Olson. 1983. Mountain Goats in the Olympic Mountain Range, Washington (1983 Census). Unpublished report on file with ONP. 17 pp.

———. 1986. An Aerial Census of Mountain Goats in the Olympic Mountain Range, Washington. *Northwest Science* 60:131–36. [P]

———. 1991. Mountain Goat Population Trends in the Olympic Mountain Range, Washington. *Northwest Science* 65:212–16. [P]

Houston, D. B., C. T. Robbins, C. A. Ruder, and R. G. Sasser. 1986. Pregnancy Detection in Mountain Goats by Assay for Pregnancy-Specific Protein B. *Journal of Wildlife Management* 50:740–42. [P*]

Houston, D. B., C. T. Robbins, and V. Stevens. 1988. Growth in Mountain Goats. *Journal of Mammalogy* 70:412–16. [P*]

Houston, D. B., and E. G. Schreiner. 1993. Biogeography of the Olympic Peninsula. Unpublished report, on file with ONP. 27 pp. [P, 1994]

Houston, D. B., E. G. Schreiner, B. B. Moorhead, and R. W. Olson. 1991. Mountain Goat Management in Olympic National Parks: A Progress Report. *Natural Areas Journal* 11:87–92. [P*]

Houston, D. B., and V. Stevens. 1988. Resource Limitation in Mountain Goats: A Test by Experimental Cropping. *Canadian Journal of Zoology* 66:228–38. [P]

Hutchins, M. 1984. The Mother–Offspring Relationships in Mountain Goats (*Oreamnos americanus*). Ph.D. dissertation, University of Washington. 350 pp.

Hutchins, M., and C. Hansen. 1982. Mother–Infant Interactions among Free-Ranging Non-native Mountain Goats in Olympic National Park. In *Ecological Research in National Parks of the Pacific Northwest,* edited by E. E. Starkey, J. F. Franklin, and J. W. Matthews, pp. 58–68. Corvallis: OSU Forest Resources Laboratory. [P*]

Hutchins, M., and V. Stevens. 1981. Olympic Mountain Goats. *Natural History* 90(1):58–69.

Interagency Goat Management Team Newsletter 1. 1992. [NOTE: tallied as published in figures 5.1 and 5.2] [P*]

Interagency Goat Management Team Newsletter 2. 1992. [NOTE: tallied as published in figures 5.1 and 5.2] [P*]

Johnson, R. L., and B. B. Moorhead. 1982. Trapping and Transplanting Mountain Goats in Washington State. In *Proceedings of the Third Biennial Symposium of the Northern Wild Sheep and Goat Council,* edited by J. A. Bailey and G. G. Schoonveld, pp. 106–14. [P*]

Jokerst, J. 1960. Population and Distribution of the Mountain Goat in ONP. Unpublished report on file with ONP. 8 pp. [U]

Kaye, T. N. 1989. Autecology, Reproductive Ecology, and Demography of *Astragalus australis* var. *olympicus* (Fabaceae). Unpublished M.S. thesis, OSU, Corvallis. 114 pp. [U 1/2, P 1/2]

Lack, L. 1962. Goat Ecology Studies, Olympic National Park. Unpublished field notes, on file with ONP. [U]

——. 1963. Goat Ecology Studies, Olympic National Park. Unpublished field notes, on file with ONP. [U]

——. 1964. Goat Ecology Studies, Olympic National Park. Unpublished field notes, on file with ONP. [U]

Luoma, J. R. 1993. Gunning for Goats. *Wildlife Conservation* 96(6): 30–41.

Machlis, G. E., S. Tuler, and R. Kasperson. 1990. A Social Risk Analysis of the Olympic National Park Mountain Goat Removal Project. University of Idaho College of Forestry, unpublished report to ONP. NPS Reference #D-173. 74 pp.

Moorhead, B. B. 1969. Ecology of Non-native Mountain Goats in the Olympic National Park. Unpublished report on file with ONP. 2 pp. [U]

——. 1971. Report of Aerial Counts of Mountain Goats. Unpublished report on file with ONP. [U]

——. 1972. Report of Aerial Counts of Mountain Goats. Unpublished report on file with ONP. [U]

——. 1973a. Report of Aerial Counts of Mountain Goats. Unpublished report on file with ONP. [U]

——. 1973b. Summary Outline of Progress. Unpublished report on file with ONP. 6 pp. [U]

——. 1974. Report of Aerial Counts of Mountain Goats. Unpublished report on file with ONP. [U]

——. 1976. Report of a 24 September 1976 Cooperative Goat Census on Mt. Angeles. Unpublished report on file with ONP. 10 pp. [Mountain Goat Ecology and Management: Results of . . . ; 11 pp.] [U]

——. 1977a. The Enigmatic Mountain Goat. *Pacific Search* 11(8):2–3.

——. 1977b. Status and Management of the Mountain Goat in the Olympic Mountains, Washington. Unpublished report on file with ONP. 16 pp. [7 pp.; 9 pp.] [U]

——. 1981a. ONP Stages Well-Run Removal of Exotic Goats. *Park Science* 2:5. [P*]

——. 1981b. Progress Report on Goat Sterilization. Unpublished report on file with ONP. 6 pp. [U]

——. 1983. Olympic National Park Mountain Goat Live-Capture and Transport Guidelines. Unpublished report on file with ONP. 11 pp. [U]

——. 1984. Transport Crate Facilitates Goat Removal. *Park Science* 4:20. [P*]

——. 1988. Non-native Mountain Goat Management Undertaken at Olympic National Park. *Park Science* 9(3):10–11. [P*]

——. 1990. An Update on Immuno-contraception and Mountain Goats. Unpublished report on file with ONP. 5 pp. [U]

Moorhead, B. B., and V. Stevens. 1982. Introduction and Dispersal of Mountain Goats in Olympic National Park. In *Ecological Research in National Parks of the Pacific Northwest*, edited by E. E. Starkey, J. F. Franklin, and J. W. Matthews, pp. 46–50. Corvallis: OSU Forest Resources Laboratory. [P]

Newman, C. C. 1954. Special Report on Mountain Goat. Unpublished report on file with ONP. 4 pp. [U]

Olmsted, I. C. 1977. Interrelationships of Introduced Mountain Goats and Subalpine Habitat in Olympic National Park. In *Proceedings of the First International Mountain Goat Symposium*, edited by W. Samuel and W. G. Macgregor, p. 200. [NOTE: this is only the abstract of the paper] [P 1/2]

——. 1979. Alpine and Subalpine Vegetation under the Influence of

Non-native Mountain Goats, Olympic National Park. In *Proceedings of the First Conference on Scientific Research in National Parks*, edited by R. M. Linn, pp. 1143–48. NPS Proceedings Series No. 5. [P]

Olson, R. W. 1984. Experimental Live Capture and Management of Interior Populations of Mountain Goats in Olympic National Park. Unpublished report on file with ONP. 42 pp. [38 pp.] [U]

——. 1992. Summary List of Mountain Goats Collected as Biological Samples or Shot in Olympic National Park, 1981–1987. Unpublished report on file with ONP. 2 pp. [U]

Olson, R. W., R. A. Hoffman, and D. E. Seaman. 1985. Selective Live Capture and Radio Telemetry of Mountain Goats in Olympic National Park, Final Report. Unpublished report on file with ONP. 35 pp. [U]

Olson, R. W., and A. T. Sandison. 1990. Modification of Hand-held Net-guns for Aerial Capture of Mountain Goats. Unpublished report on file with ONP. 8 pp. [U]

Olympic National Park. 1981. An Environmental Assessment on the Management of Introduced Mountain Goats in Olympic National Park. Unpublished ONP report. 49 pp. [U]

——. 1982. Mountain Goat Ecology and Management Investigations, Olympic National Park, 1981. Unpublished ONP report. 68 pp. [U]

——. 1983a. Backcountry Experimental Goat Management Project. Unpublished report on file with ONP. 27 pp. [U]

——. 1983b. Mountain Goat Ecology and Management Investigations, Olympic National Park, 1982. Unpublished ONP report. 65 pp. [U]

——. 1984. Mountain Goat Ecology and Management Investigations, Olympic National Park, 1983. Unpublished ONP report. 142 pp. [U]

——. 1985. Draft Status Report: Experimental Goat Management. Unpublished report on file with ONP. 9 pp. [U]

——. 1986. Goat Management Program Summary. Unpublished report on file with ONP. 12 pp. [U]

——. 1987a. The Management of Mountain Goats in Olympic National Park: Some Questions and Some Answers. Flier distributed by ONP. 9 pp. [NOTE: tallied as published in figures 5.1 and 5.2] [P*]

——. 1987b. Mountain Goat Management in Olympic National Park. Unpublished report on file with ONP. 72 pp. [U]

——. 1989. Statement for Management. Unpublished report on file with ONP. [U]

——. 1992a. Addendum to Backcountry Management Plan. Unpublished report on file with ONP. [U]

——. 1992b. Resources Management Plan. Unpublished report on file with ONP. [U]

Pfitsch, W. A. 1981. The Effect of Mountain Goats on the Subalpine Plant Communities of Klahhane Ridge, Olympic National Park. M.S. thesis, UW. 113 pp.

Pfitsch, W. A., and L. C. Bliss. 1985. Seasonal Forage Availability and Potential Vegetation Limitations to a Mountain Goat Population, Olympic National Park. *American Midland Naturalist* 113:109–21. [P 1/2; because it does not contain all of Pfitsch's (1981), Pike's (1981), and Reid's (1983) data]

Pfitsch, W. A., R. S. Reid, J. Harter, D. K. Pike, and L. C. Bliss (eds.). 1983. Effects of Mountain Goats on Soils, Plant Communities and Selected Species in Olympic National Park. UW Department of Botany unpublished report, on file with ONP. 105 pp.

Pike, D. K. 1981. Effects of Mountain Goats on Three Plant Species Unique to the Olympic Mountains, Washington. M.S. thesis, UW. 188 pp.

Reid, R. S. 1983. The Effect of Mountain Goats on Plant Establishment and Seedling Survival, Olympic National Park. [Patterns of Juvenile Mortality and Plant Life Histories in Response to Mountain Goat Disturbance, Olympic National Park, Washington] M.S. thesis, UW. 215 pp.

Schalk, R. F. 1993. A Review of the Ethnographic and Archaeological Evidence Relating to Mountain Goats in the Olympic Mountains. Unpublished report on file with ONP. 60 pp. [U]

Scheffer, V. B. 1993. The Olympic Mountain Goat Controversy: A Perspective. *Conservation Biology* 7:916–19.

Schreiner, E. G. 1984. Soils and Vegetation (Draft). Unpublished report on file with ONP. 7 pp. [U]

——. 1986. Vegetation Changes between 1979 and 1985 on Klahhane Ridge. Unpublished report on file with ONP. 9 pp. [U]

——. 1987. Vegetation in Relation to Nonnative Mountain Goats in the

Olympic Mountains, Washington. Unpublished report on file with ONP. 111 pp. [U]

——. 1989. Monitoring/Study of the Relationships between Non-native Mountain Goats and Vegetation in Olympic National Park: A Summary and Proposal. Unpublished report on file with ONP. 4 pp. [U]

Schreiner, E. (G.), A. Woodward, and M. Gracz. 1993. Vegetation in Relation to Introduced Mountain Goats in Olympic National Park: A Technical Report. Unpublished report on file with ONP. 154 pp. [P, 1994]

Schultz, S. 1993. A Review of the Historical Evidence Relating to Mountain Goats in the Olympic Mountains Prior to 1925. Unpublished report on file with ONP. 75 pp. [P, 1994]

Sleeper, B. 1990. Out on a Ledge. *Animals* 23(4):18–25.

Sleeper, B. L., and M. Hutchins. 1989. The Horns of a Dilemma. *Pacific Discovery* 42(1):16–24.

Stevens, V. 1977. Unpublished NPS progress report. On file with ONP. [U]

——. 1978a. Progress Report, June 12–July 19, 1978. Unpublished report on file with ONP. 5 pp. [U]

——. 1978b. Progress report, July 20–September 6, 1978. Unpublished report on file with ONP. 12 pp. [U]

——. 1979a. Mountain Goat Habitat Utilization in Olympic National Park. M.S. thesis, UW. 106 pp. [U]

——. 1979b. Progress Report, June 3–July 25, 1979. Unpublished report on file with ONP. 19 pp. [U]

——. 1979c. Progress Report, July 30–September 22, 1979. Unpublished report on file with ONP. 13 pp. [U]

——. 1979d. Terrestrial Baseline Surveys, Non-native Mountain Goats of the Olympic National Park. UW College of Forest Resources, unpublished report on file with ONP. 77 pp. [Stevens, Driver, and Taber: Annual Report: Terrestrial Baseline . . . ; 75 pp.] [U]

——. 1980. Final Report, Terrestrial Baseline Surveys, Non-native Mountain Goats of the Olympic National Park. [Taber and Stevens: Terrestrial Baseline Surveys . . . ; Stevens and Taber: Final Report . . .] UW College of Forest Resources, unpublished report on file with ONP. 145 pp. [U]

——. 1982. Factors Reflecting Mountain Goat Condition and Habitat

Quality: A Comparison of Subpopulations in Olympic National Park. In *Ecological Research in National Parks of the Pacific Northwest*, edited by E. E. Starkey, J. F. Franklin, and J. W. Matthews, pp. 51–57. Corvallis: OSU Forest Resources Laboratory. [P]

——. 1983. Dynamics of Dispersal in an Introduced Mountain Goat Population on the Olympic Peninsula, Washington. Ph.D. dissertation, UW. 202 pp. [U]

Stevens, V., and C. Driver. 1978. Initial Observations on a Tagged Mountain Goat Population in the Olympic Mountains. In *Proceedings of the Biennial Symposium of the Northern Wild Sheep and Goat Council*, edited by D. M. Hebert and M. Nation, pp. 165–74. [P]

Stevens, V., and D. B. Houston. 1989. Reliability of Age Determination of Mountain Goats. *Wildlife Society Bulletin* 17:72–74. [P*]

Stevens, V., and R. D. Taber. 1981. Supplemental Report: Terrestrial Baseline Surveys, Non-native Mountain Goats of the Olympic National Park. UW College of Forest Resources, unpublished report on file with ONP. 25 pp. [U]

Tuler, S., and C. Janda. 1991. Olympic NP Mountain Goat Removal Project Subject of Risk Assessment Report. *Park Science* 11(1):14–15.

Tuler, S., G. E. Machlis, and R. E. Kasperson. 1992. Mountain Goat Removal in Olympic National Park: A Case Study of the Role of Organizational Culture in Individual Risk Decisions and Behavior. *RISK* 3(4):317–40. [P]

USDA Forest Service. 1954. Washington Mountain Goat Population—1954. Unpublished report on file with ONP. 1 p.

Warren, R. J. 1993. Report to ONP Superintendent. Unpublished report on file with ONP. 3 pp. [U]

Warren, R. J., R. A. Fayrer-Hosken, D. A. Jessup, R. A. Garrott, and J. F. Kirkpatrick. 1992. The Applicability of Contraceptives in the Elimination or Control of Exotic Mountain Goats from Olympic National Park. Unpublished report on file with ONP. 35 pp. [U]

REFERENCES ON WAPITI

Bellingham, D. 1971. Elk Movement and Distribution. Job Completion Report, Project No. W-75-R-3, Washington Department of Wildlife, Olympic, Wash. 7 pp.

Brent, H. 1957. Quarterly Narrative Report for Investigation Projects:

Big Game Research, Project No. W-37-R-9. USDA Forest Service, Quinault.

Hanley, T. A., and R. D. Taber. 1980. Selective Plant Species Inhibition by Elk and Deer in Three Conifer Communities in Western Washington. *Forestry Science* 26:97–107.

Happe, P. J. 1993. Elk-Vegetation Relationships on the South Fork Hoh River, Washington. Ph.D. dissertation, OSU.

Harmon, M. E., and J. F. Franklin. 1983. Age Distribution of Western Hemlock and Its Relation to Roosevelt Elk Populations in the South Fork Hoh River Valley, Washington. *Northwest Science* 57:249–55.

Harrison, R. H. 1963. Quinault District Elk Exclosures. Unpublished report, USDA Forest Service, Quinault. 10 pp.

Houston, D. B., B. B. Moorhead, and R. W. Olson. 1987. Roosevelt Elk Density in Old-Growth Forests of Olympic National Park. *Northwest Science* 61:220–25.

——. 1988. Roosevelt Elk Census in Olympic National Park: 1984–1988. Unpublished report on file with ONP. 8 pp.

Houston, D. B., E. G. Schreiner, B. B. Moorhead, and K. A. Krueger. 1990. Elk in the Olympic National Park: Will They Persist over Time? *Natural Areas Journal* 10:6–11.

Jenkins, K. J. 1979. Home Ranges and Habitat Use by Roosevelt Elk in Olympic National Park, WA. M.S. thesis, OSU.

——. 1981. Status of Elk Populations and Lowland Habitats in Western Olympic National Park. Cooperative Park Studies Unit, OSU, Report 81-1. 53 pp.

Jenkins, K. J., and E. E. Starkey. 1980a. Home Range and Habitat Use by Non-migratory Roosevelt Elk, Olympic National Park. In *Ecological Research in National Parks of the Pacific Northwest,* edited by E. E. Starkey, J. F. Franklin, and J. W. Matthews, pp. 69–76. Corvallis: OSU Forest Resources Laboratory.

——. 1980b. Roosevelt Elk of the Hoh Valley, Olympic National Park. Cooperative Park Studies Unit Report, OSU, Report 80-3. 32 pp.

——. 1982. Social Organization of Roosevelt Elk in an Old-Growth Forest. *Journal of Mammalogy* 63:331–34.

——. 1984. Habitat Use by Roosevelt Elk in Unmanaged Forests of the Hoh Valley, Washington. *Journal of Wildlife Management* 48: 642–46.

Leslie, D. M. 1983. Nutritional Ecology of Cervids in Old-Growth Forests in Olympic National Park, WA. Ph.D. dissertation, OSU.

Leslie, D. M., and E. E. Starkey. 1982. Cervid-Habitat Interactions in Olympic National Park. OSU School of Forestry Report 82-2, OSU. 109 pp.

Leslie, D. M., E. E. Starkey, and M. Vavra. 1984. Elk and Deer Diets in Old-Growth Forests in Western Washington. *Journal of Wildlife Management* 48:762–75.

Leslie, D. M., M. Vavra, E. E. Starkey, and R. C. Slater. 1983. Correcting for Differential Digestibility in Microhistological Analysis Involving Common Coastal Foragers of the Pacific Northwest. *Journal of Range Management* 36:731–32.

Newman, C. 1953. Report on Roosevelt Elk in Olympic National Park. Unpublished report on file with ONP.

———. 1954. Report on Roosevelt Elk in Olympic National Park. Unpublished report on file with ONP.

———. 1955. Report on Roosevelt Elk in Olympic National Park. Unpublished report on file with ONP.

———. 1956. Biologist's Annual Report—Elk. Unpublished report on file with ONP.

———. 1957. Report on Roosevelt Elk in Olympic National Park. Unpublished report on file with ONP.

———. 1958. Report on Roosevelt Elk in Olympic National Park. Unpublished report on file with ONP.

Pfister, G., and B. Pfister. 1989. Quinault Valley Elk Exclosures of Olympic National Park. Unpublished report on file with ONP.

Schroer, G. L. 1987. Seasonal Movements and Distribution of Migratory Roosevelt Elk in the Olympic Mountains, Washington. M.S. thesis, OSU. 86 pp.

Schroer, G. L., K. J. Jenkins, and B. B. Moorhead. 1993. Roosevelt Elk Selection of Temperate Rain Forest Seral Stages in Western Washington. *Northwest Science* 67:23–29.

Schwartz, J. E. 1939. The Olympic Elk Study. Unpublished report on file with US Forest Service, Olympia.

Schwartz, J. E., and G. E. Mitchell. 1945. The Roosevelt Elk on the Olympic Peninsula. *Journal of Wildlife Management* 9:295–319.

Skinner, M. P. 1933. Report on Elk Conditions in Olympic Mountains to Boone and Crockett Club. Unpublished report on file with ONP.

——. 1936a. Browsing of the Olympic Peninsula Elk in Early Winter. *Journal of Mammalogy* 17:253–56.

——. 1936b. The Roosevelt Elk of the Olympic Peninsula. *Murrelet* 17(2–3):31–35.

Smith, J. R. 1980. Reproductive Rates, Age Structure and Management of Roosevelt Elk in Washington's Olympic Mountains. In *Proceedings of the Western States Elk Workshop*, edited by W. Macgregor, pp. 67–111. British Columbia Fish and Game Branch, Victoria.

Sumner, E. L. 1938. Special Report of Elk in Olympic National Park. Unpublished report on file with NPS, Washington, D.C. 18 pp.

Taber, R. D., and K. J. Raedeke. 1980. Roosevelt Elk of the Olympic National Forest. UW College of Forest Resource, unpublished report. 107 pp.

Woodward, A., E. G. Schreiner, D. B. Houston, and B. B. Moorhead. 1993. Ungulate-Forest relationships in Olympic National Park: Retrospective Exclosure Studies. USDI, NPS Regional Technical Report NPS/PNROLYM/NRTR-93/15, PNW Region, Seattle. 62 pp.

REFERENCES

Ackerman, Edward A., Martson Bates, Stanley A. Cain, F. Fraser Darling, John M. Fogg, Jr., Tom Gill, Joseph M. Gillson, E. Raymond Hall, Carl L. Hubbs, William J. Robbins, and C. J. S. Durham. 1963. *A Report by the Advisory Committee to the National Park Service on Research.* Washington, D.C.: National Academy of Science, Natural Resources Council.

Alien to Olympic Park, Mountain Goats May Go. 1992. *National Geographic* 181(6):136.

Anderson, Jay E. 1991. A Conceptual Framework for Evaluating and Quantifying Naturalness. *Conservation Biology* 5:347–52.

Anunsen, Cathy Sue, and Roger Anunsen. 1993. Response to Scheffer. *Conservation Biology* 7:954–57.

Armstrong, David M. 1972. *Distribution of Mammals in Colorado.* Monograph 3. Lawrence: University of Kansas Museum of Natural History.

Bailey, James A. 1982. Implications of "Muddling Through" for Wildlife Management. *Wildlife Society Bulletin* 10:363–69.

Bailey, Vernon. 1931. The Mountain Goat That Is Not a Goat. *Nature Magazine* 18:153–57.

——. 1936. *The Mammals and Lifezones of Oregon.* North American Fauna 55:1–416.

Bambach, Richard K., and J. Bret Bennington. 1996. Do Communities Evolve? A Major Question in Evolutionary Paleobiology. In *Evolutionary Paleobiology,* edited by David Jablonski, Douglas H. Erwin, and Jere H. Lipps, pp. 123–60. Chicago: University of Chicago Press.

Banfield, A. W. F. 1974. *The Mammals of Canada.* Toronto: National Museums of Canada, University of Toronto Press.

Barber, Bernard. 1961. Resistance by Scientists to Scientific Discovery. *Science* 134:596–602.

Barnes, C. A. 1890. Found in the Olympics: A Resume of the Natural Resources of Explored Region. *Seattle Press*, July 16.

Barnes, Lawrence G., and James L. Goedert. 1996. Marine Vertebrate Paleontology on the Olympic Peninsula. *Washington Geology* 24(3): 17–25.

Barnosky, Cathy W. 1984. Late Pleistocene and Early Holocene Environmental History of Southwestern Washington State, U.S.A. *Canadian Journal of Earth Sciences* 21:619–29.

Barnosky, Cathy W., Patricia M. Anderson, and Patrick J. Barlein. 1987. The Northwestern U.S. during Deglaciation; Vegetational History and Paleoclimatic Inference. In *North America and Adjacent Oceans during the Last Deglaciation*, edited by W. F. Ruddiman and H. E. Wright, Jr., pp. 289–321. Geology of North America vol. K-3. Boulder: Geological Society of America.

Bartels, Molly A., and Doug P. Thompson. 1993. *Spermophilus lateralis. Mammalian Species* 440:1–8.

Bell, Katherine L., and Lawrence C. Bliss. 1973. Alpine Disturbance Studies Olympic National Park, USA. *Biological Conservation* 5:25–32.

Bennett, Dean. 1994. The Unique Contribution of Wilderness to Values of Nature. *Natural Areas Journal* 14:203–8.

Berger, Joel. 1990. Persistence of Different-Sized Populations: An Empirical Assessment of Rapid Extinctions in Bighorn Sheep. *Conservation Biology* 4:91–98.

Bergland, Eric O., and Jerry Marr. 1988. *Prehistoric Life on the Olympic Peninsula: The First Inhabitants of a Great American Wilderness.* Seattle: Pacific Northwest National Parks and Forests Association.

Between a Rock and a Hard Place. 1995. *Sunset* 195(1):133.

Bishop, Roy, and Rita Martin. 1994. Olympic Goats. *National Parks* 68(9–10):8.

Bleich, V. C., J. D. Wehausen, and S. A. Holl. 1990. Desert-Dwelling Mountain Sheep: Conservation Implications of a Naturally Fragmented Distribution. *Conservation Biology* 4:383–90.

Bodmer, Richard E., John F. Eisenberg, and Kent H. Redford. 1997. Hunting and the Likelihood of Extinction of Amazonian Mammals. *Conservation Biology* 11:460–66.

Bomford, Mary, and Peter O'Brien. 1995. Eradication or Control for Vertebrate Pests? *Wildlife Society Bulletin* 23:249–55.

Bonnicksen, Thomas M., and Edward C. Stone. 1982. Managing Vegetation within U.S. National Parks: A Policy Analysis. *Environmental Management* 6:101–2, 109–22.

——. 1985. Restoring Naturalness to National Parks. *Environmental Management* 9:479–86.

Booth, Derek B. 1987. Timing and Processes of Deglaciation along the Southern Margin of the Cordilleran Ice Sheet. In *North America and Adjacent Oceans during the Last Deglaciation*, edited by W. F. Ruddiman and H. E. Wright, Jr., pp. 71–90. Geology of North America vol. K-3. Boulder: Geological Society of America.

Booth, Ernest S. 1947. Systematic Review of the Land Mammals of Washington. Ph.D. dissertation, State College of Washington (Washington State University), Pullman.

Boyd, Robert T. 1990. Demographic History, 1774–1874. In *Handbook of North American Indians Vol. 7: Northwest Coast*, edited by Wayne Suttles, pp. 135–48. Washington, D.C.: Smithsonian Institution Press.

Bradley, William P., and Chas. H. Driver. 1981. Elk Ecology and Management Perspectives at Mount Rainier National Park. University of Washington, Cooperative Park Studies Unit, Report 81-2.

Brandborg, Stewart M. 1955. *Life History and Management of the Mountain Goat in Idaho.* Idaho Department of Fish and Game Wildlife Bulletin No. 2.

Brett, Carlton E., Linda C. Ivany, and Kenneth M. Schopf. 1996. Coordinated Stasis: An Overview. *Palaeogeography, Palaeoclimatology, Palaeoecology* 127:1–20.

Brooks, Allan. 1930. Early Big-Game Conditions in the Mount Baker District, Washington. *Murrelet* 11:65–67.

Brown, James H. 1971. Mammals on Mountaintops: Nonequilibrium Insular Biogeography. *American Naturalist* 105:467–78.

——. 1978. The Theory of Insular Biogeography and the Distribution of Boreal Birds and Mammals. In *Intermountain Biogeography: A Symposium*, edited by K. T. Harper and James L. Reveal, pp. 209–27. Great Basin Naturalist Memoirs 2.

——. 1986. Two Decades of Interaction between the MacArthur–Wilson Model and the Complexities of Mammalian Distributions. In *Is-*

land Biogeography of Mammals, edited by L. R. Heaney and Bruce D. Patterson, pp. 231–51. Biological Journal of the Linnean Society 28.

Brown, James H., and Arthur C. Gibson. 1983. *Biogeography*. St. Louis: C. V. Mosby Company.

Brown, James H., and Brian A. Maurer. 1989. Macroecology: The Division of Food and Space among Species on Continents. *Science* 243:1145–50.

Brown, James H., and Don E. Wilson. 1994. Natural History and Evolutionary Ecology. In *Seventy-five Years of Mammalogy (1919–1994)*, edited by Elmer C. Birney and Jerry R. Choate, pp. 377–97. American Society of Mammalogists Special Publication No. 11.

Browne, Belmore. 1910. The First Ascent of Mount Olympus. *Bulletin of the American Geographical Society* 42:881–87.

Brush, Stephen G. 1974. Should the History of Science Be Rated X? *Science* 183:1164–72.

Buechner, Helmut K. 1953. Some Biotic Changes in the State of Washington, Particularly during the Century 1853–1953. *Research Studies of the State College of Washington* 21:154–92.

Burger, Janis. 1987. A Mountainous Appetite. *National Parks* 61(1–2): 28–31.

Burns, James A. 1987. Late Quaternary Zoogeography of the Northern Pocket Gopher, *Thomomys talpoides*, in Southwestern Alberta. *Canadian Field-Naturalist* 101:419–22.

Busch, Lisa. 1994. A Glimmer of Hope for Coastal Migration. *Science* 263:1088–89.

Campbell, Erick G., and Rolf L. Johnson. 1983. Food Habits of Mountain Goats, Mule Deer and Cattle on Chopaka Mountain, Washington, 1977–1980. *Journal of Range Management* 36:488–91.

Cannon, Aubrey. 1991. *The Economic Prehistory of Namu: Patterns in Vertebrate Fauna.* Department of Archaeology Publication No. 19. Burnaby, B.C.: Simon Fraser University.

Carl, G. Clifford, and C. J. Guiguet. 1972. *Alien Animals in British Columbia.* Victoria, B.C.: British Columbia Provincial Museum Handbook 14.

Carlquist, Brad. 1990. An Effective Management Plan for the Exotic Mountain Goats in Olympic National Park. *Natural Areas Journal* 10:12–18.

Carraway, Leslie N., Lois F. Alexander, and B. J. Verts. 1993. *Scapanus townsendii. Mammalian Species* 434:1–7.

Chadwick, Douglas H. 1983. *A Beast the Color of Winter: The Mountain Goat Observed.* San Francisco: Sierra Club Books.

Chamberlain, T. C. 1897. The Method of Multiple Working Hypotheses. *Journal of Geology* 39:155–65.

Chase, Alston. 1987. *Playing God in Yellowstone: The Destruction of America's First National Park.* 2nd ed. San Diego: Harcourt Brace Jovanovich.

Cheetham, Alan H., and Joseph E. Hazel. 1969. Binary (Presence–Absence) Similarity Coefficients. *Journal of Paleontology* 43: 1130–36.

Clarke, Jennifer A., and Richard A. Johnson. 1990. Biogeography of White-Tailed Ptarmigan (*Lagopus leucurus*): Implications from an Introduced Population in the Sierra Nevada. *Journal of Biogeography* 17:649–56.

Coblentz, Bruce E. 1990. Exotic Organisms: A Dilemma for Conservation Biology. *Conservation Biology* 4:261–65.

Collier, Donald, Alfred E. Hudson, and Arlo Ford. 1942. *Archaeology of the Upper Columbia Region.* University of Washington Publications in Anthropology 9:1–178.

Colwell, Robert K. 1992. Niche: A Bifurcation in the Conceptual Lineage of the Term. In *Keywords in Evolutionary Biology,* edited by Evelyn Fox Keller and Elisabeth A. Lloyd, pp. 241–48. Cambridge, Mass.: Harvard University Press.

Conover, Kathryn. 1978. Matrix Analysis. In *Studies in Bella Bella Prehistory,* edited by James J. Hester and Sarah M. Nelson, pp. 67–99. Department of Archaeology Publication No. 5. Burnaby, B.C.: Simon Fraser University.

Cook, Rosamonde R., and James F. Quinn. 1995. The Influence of Colonization in Nested Species Subsets. *Oecologia* 102:413–24.

Coté, Steve D. 1996. Mountain Goat Responses to Helicopter Disturbance. *Wildlife Society Bulletin* 24:681–85.

Cowan, Ian McTaggart, and Charles J. Guiguet. 1965. *The Mammals of British Columbia.* Victoria, B.C.: British Columbia Provincial Museum Handbook 11.

Cowan, Ian McTaggart, and W. McCrory. 1970. Variation in the Mountain Goat, *Oreamnos americanus* (Blainville). *Journal of Mammalogy* 51:60–73.

Crawford, Paul. 1993. Olympic Mountain Goat Update. *Park Science* 13(3):15.

Crockford, Susan. 1994. Osteometric and Ancient DNA Analysis of Prehistoric Dogs of the Central Northwest Coast: Wool Dog or Bust! *Canadian Zooarchaeology* 5:15–20.

Cronin, Matthew A. 1993. Mitochondrial DNA in Wildlife Taxonomy and Conservation Biology: Cautionary Notes. *Wildlife Society Bulletin* 21:339–48.

Curtin, Charles G. 1995. Can Montane Landscapes Recover from Human Disturbance? Long-term Evidence from Disturbed Subalpine Communities. *Biological Conservation* 74:49–55.

Curtis, Asahel. 1908. The First Ascent of Mt. Olympus. *World's Work* 16:10261–62.

Cutler, Alan. 1991. Nested Faunas and Extinction in Fragmented Habitats. *Conservation Biology* 5:496–505.

Dalquest, Walter W. 1948. *Mammals of Washington.* University of Kansas Museum of Natural History Publications (Lawrence) 2:1–444.

Daugherty, Richard D., J. Jeffrey Flenniken, and Jeanne M. Welch. 1987a. *A Data Recovery Study of Judd Peak Rockshelters (45LE222) in Lewis County, Washington.* Studies in Cultural Resource Management No. 8. Portland, Ore.: USDA Forest Service, Pacific Northwest Region.

———. 1987b. *A Data Recovery Study of Layser Cave (45LE223) in Lewis County, Washington.* Studies in Cultural Resource Management No. 7. Portland, Ore.: USDA Forest Service, Pacific Northwest Region.

Davis, David E., and Ray L. Winstead. 1980. Estimating the Numbers of Wildlife Populations. In *Wildlife Management Techniques Manual,* edited by Sanford D. Schemnitz, pp. 221–45. 4th ed. Washington, D.C.: Wildlife Society.

Davis, Gary E., and William L. Halvorson. 1996. Long-term Research in National Parks: From Beliefs to Knowledge. In *Science and Ecosystem Management in the National Parks,* edited by William L. Halvorson and Gary E. Davis, pp. 3–10. Tucson: University of Arizona Press.

Davis, Russell, Christopher Dunford, and Mark V. Lomolino. 1988. Montane Mammals of the American Southwest: The Possible Influence of Post-Pleistocene Colonization. *Journal of Biogeography* 15:841–48.

Davis, William B. 1939. *The Recent Mammals of Idaho.* Caldwell, Idaho: Caxton Printers.

Del Moral, Roger. 1984. The Impact of the Olympic Marmot on Subalpine Vegetation Structure. *American Journal of Botany* 71:1228–36.

Denevan, William M. 1992. The Pristine Myth: The Landscape of the Americas in 1492. *Annals of the Association of American Geographers* 82:369–85.

Dice, Lee R. 1932. Mammals Collected by F. M. Gaige in 1919 at Lake Cushman and Vicinity, Olympic Peninsula, Washington. *Murrelet* 13:47–49.

Dilsaver, Lary M. (ed.). 1994. *America's National Park System: The Critical Documents.* Lanham, Md.: Rowman and Littlefield.

Dixon, C., and B. Leach. 1978. *Sampling Methods for Geographical Research.* Concepts and Techniques in Modern Geography No. 17. London: Institute of British Geographers.

Dixon, Susan. 1996. Historical Faunal Turnover: Replicable Documentation of Change in Mount Rainier National Park. *MU McNair Journal* 4:4–9.

Doak, Daniel F., and L. Scott Mills. 1994. A Useful Role for Theory in Conservation. *Ecology* 75:615–26.

Dorrance, Michael J. 1983. A Philosophy of Problem Wildlife Management. *Wildlife Society Bulletin* 11:319–24.

Draft Guidelines for Reintroductions. 1992. *Newsletter of the IUCN Reintroduction Specialist Group* 4:2–3.

Dunlap, Thomas R. 1988. *Saving America's Wildlife.* Princeton: Princeton University Press.

Dunnell, Robert C. 1991. Methodological Impacts of Catastrophic Depopulation on American Archaeology and Ethnology. In *Columbian Consequences, Vol. 3: The Spanish Borderlands in Pan-American Perspective,* edited by David Hurst Thomas, pp. 561–80. Washington, D.C.: Smithsonian Institution Press.

Dunwiddie, Peter W. 1986. A 6000-Year Record of Forest History on Mount Rainier, Washington. *Ecology* 67:58–68.

Easterbrook, Don J. 1986. Stratigraphy and Chronology of Quaternary Deposits of the Puget Lowland and Olympic Mountains of Washington and the Cascade Mountains of Washington and Oregon. *Quaternary Science Reviews* 5:145–59.

———. 1992. Advance and Retreat of Cordilleran Ice Sheets in Washington, U.S.A. *Géographie Physique et Quaternaire* 46:51–68.

Edson, J. M. 1930. Wild Animal Populations of the Mount Baker National Forest. *Murrelet* 11:14–15.

———. 1931. Again the Big-Game of the Mount Baker District, Washington. *Murrelet* 12:50–53.

Elliot, D. G. 1899. Catalogue of Mammals from the Olympic Mountains, Washington, with Descriptions of New Species. *Field Columbian Museum Publication* 32, *Zoological Series* 1(13):239–76.

Elmendorf, William W. 1960. *Structure of Twana Culture*. Research Studies, Monograph Supplement No. 2. Pullman: Washington State University.

——. 1993. *Twana Narratives: Native Historical Accounts of a Coast Salish Culture*. Seattle: University of Washington Press.

Engstrom, Mark D., Jerry R. Choate, and Hugh H. Genoways. 1994. Taxonomy. In *Seventy-Five Years of Mammalogy (1919–1994)*, edited by Elmer C. Birney and Jerry R. Choate, pp. 179–199. American Society of Mammalogists Special Publication No. 11.

Evans, Martin. 1997. Pollen Evidence of Late Holocene Treeline Fluctuation from the Southern Coast Mountains, British Columbia. *Géographie Physique et Quaternaire* 51:81–92.

Fannin, John. 1890. The Rocky Mountain Goat. In *The Big Game of North America*, edited by G. O. Shields, pp. 343–61. Chicago: Rand McNally.

Fannin, John, and George Bird Grinnell. 1890a. Chase of the White Goat. *Forest and Stream*, August 21, p. 88.

——. 1890b. Habits of the White Goat. *Forest and Stream*, July 31, pp. 27–28; August 7, p. 47; August 14, p. 67.

——. 1890c. Range of the White Goat. *Forest and Stream*, February 18, vol. 34:62–64.

Findley, James S. 1969. Biogeography of Southwestern Boreal and Desert Mammals. *University of Kansas Museum of Natural History Miscellaneous Publications* 51:113–28.

Fischer, D. H. 1970. *Historians' Fallacies: Toward a Logic of Historical Thought*. New York: Harper and Row.

Fischer, Hank. 1995. *Wolf Wars: The Remarkable Inside Story of the Restoration of Wolves to Yellowstone*. Helena, Mont.: Falcon Press.

Flanagan, Joseph. 1995. Locking Horns with History. *Federal Archaeology* 8(1):22–26.

Foley, Patrick. 1994. Predicting Extinction Times from Environmental Stochasticity and Carrying Capacity. *Conservation Biology* 8:124–37.

Fox, Joseph L., and Gregory P. Streveler. 1986. Wolf Predation on Mountain Goats in Southeastern Alaska. *Journal of Mammalogy* 67: 192–95.

Frankham, Richard. 1996. Relationships of Genetic Variation to Population Size in Wildlife. *Conservation Biology* 10:1500–1508.

Franklin, Jerry F. 1977. The Biosphere Reserve Program in the United States. *Science* 195:262–67.

Frasier, Gary W. 1988. Peer Review of Technical Manuscripts. *Journal of Range Management* 41:366–67.

Freed, J. Arthur, and Kenneth S. Lane. 1964. Analysis of Faunal Remains from Old Town, Knight Island. In *Archeology of the Yakutat Bay Area, Alaska*, by Frederica de Laguna, Francis A. Riddell, Donald F. McGeein, Kenneth S. Lane, and J. Arthur Freed, pp. 77–84. Bureau of American Ethnology Bulletin 192.

Furlong, E. L. 1906. The Exploration of Samwel Cave. *American Journal of Science* 22:235–47.

Furtman, Michael. 1993. Malice Doesn't Live Here Anymore. *Outdoor America* 58(4):11–14.

Galdikas-Brindamour, Birute. 1972. Faunal Material from Eight Archaeological Sites: A Preliminary Report. In *Salvage '71: Reports on Salvage Archaeology Undertaken in British Columbia in 1971*, edited by Roy L. Carlson, pp. 199–205. Department of Archaeology Publication No. 1. Burnaby: Simon Fraser University.

Gallison, James D. 1991. Slab Camp: An Early to Mid Holocene Occupation in the Northeastern Montane Zone of the Olympic Mountains, Quilcene District, Olympic National Forest. Unpublished draft report on file, Olympic National Forest, Olympia.

——. 1994. Slab Camp: An Early to Middle Holocene Olcott Complex in the Eastern Olympic Mountains. Ph.D. dissertation, Washington State University, Pullman.

Gaston, Kevin J. 1991. How Large Is a Species' Geographic Range? *Oikos* 61:434–38.

Gaston, Kevin J., and Tim M. Blackburn. 1996. Conservation Implications of Geographic Range Size–Body Size Relationships. *Conservation Biology* 10:638–46.

Gavin, Thomas A. 1989. What's Wrong with the Questions We Ask in Wildlife Research? *Wildlife Society Bulletin* 17:345–50.

——. 1991. Why Ask "Why": The Importance of Evolutionary Biology in Wildlife Science. *Journal of Wildlife Management* 55:760–66.

Gibbs, George. 1877. Tribes of Western Washington and Northwestern Oregon. In *Contributions to North American Ethnology* 1(2), edited by John Wesley Powell, pp. 157–361. U.S. Geographical and Geological Survey of the Rocky Mountain Region, Washington. Washington, D.C.: U.S. Government Printing Office.

Gilbow, Delbert W. 1981. Inference of Human Activity from Faunal Remains. M.A. thesis, Washington State University, Pullman.

Gilman, S. C. 1896. The Olympic Country. *National Geographic* 7:133–40.

Gleeson, Paul F. 1970. Dog Remains from the Ozette Village Archaeological Site. M.A. thesis, Washington State University, Pullman.

Glenn, Susan M., and Tom D. Nudds. 1989. Insular Biogeography of Mammals in Canadian Parks. *Journal of Biogeography* 16:261–68.

Grabert, Garland F. 1974. Okanogan Archaeology: 1966–1967. *Syesis* 7, Supplement 2:1–82.

Graham, Russell W. 1988. The Role of Climatic Change in the Design of Biological Reserves: The Paleoecological Perspective for Conservation Biology. *Conservation Biology* 2:391–94.

Grant, Madison. 1905. The Rocky Mountain Goat. *Annual Report of the New York Zoological Society* 9:230–61.

Gray, Gary G. 1993. *Wildlife and People: The Human Dimensions of Wildlife Ecology*. Urbana: University of Illinois Press.

Grayson, Donald K. 1977. On the Holocene History of Some Great Basin Lagomorphs. *Journal of Mammalogy* 58:507–513.

——. 1981. A Mid-Holocene Record for the Heather Vole, *Phenacomys* cf. *intermedius*, in the Central Great Basin, and Its Biogeographic Significance. *Journal of Mammalogy* 62:115–21.

——. 1982. Toward a History of Great Basin Mammals during the Past 15,000 Years. In *Man and Environment in the Great Basin*, edited by David B. Madsen and James F. O'Connell, pp. 82–101. Society for American Archaeology Papers No. 2.

——. 1987. The Biogeographic History of Small Mammals in the Great Basin: Observations on the Last 20,000 Years. *Journal of Mammalogy* 68:359–75.

——. 1993. *The Desert's Past: A Natural Prehistory of the Great Basin.* Washington, D.C.: Smithsonian Institution Press.

Grayson, Donald K., and Stephanie D. Livingston. 1993. Missing Mammals on Great Basin Mountains: Holocene Extinctions and Inadequate Knowledge. *Conservation Biology* 7:527–32.

Grayson, Donald K., Stephanie D. Livingston, Eric Rickart, and Monson W. Shaver III. 1996. Biogeographic Significance of Low-Elevation

Records for *Neotoma cinerea* from the Northern Bonneville Basin, Utah. *Great Basin Naturalist* 56:191–96.

Grinnell, Hilda W. 1943. Bibliography of Clinton Hart Merriam. *Journal of Mammalogy* 24:436–57.

Gunther, Erna. 1927. Klallam Ethnography. *University of Washington Publications in Anthropology* 1(5):171–314.

Gurfinkel, D. M., and U. M. Franklin. 1988. A Study of the Feasibility of Detecting Blood Residue on Artifacts. *Journal of Archaeological Science* 15:83–97.

Gustafson, Carl E. 1983. Wapiti Populations in and Adjacent to Mount Rainier National Park: Archaeological and Ethnographic Evidence. Unpublished report on file, Mount Rainier National Park, Longmire, Washington.

———. 1985. The Manis Mastodon Site. *National Geographic Research Reports* 20:283–95.

———. n.d.a. The Manis Mastodon Site. Undated, unpaginated pamphlet.

———. n.d.b. The Manis Mastodon Site: An Adventure in Prehistory. Undated, unpaginated pamphlet.

Gustafson, Carl E., Delbert Gilbow, and Richard D. Daugherty. 1979. The Manis Mastodon Site: Early Man on the Olympic Peninsula. *Canadian Journal of Archaeology* 3:157–64.

Gutterson, David. 1993. Warm, Fuzzy and Unwelcome. *Sports Illustrated* 78(13).

Haeberlin, Hermann, and Erna Gunther. 1930. The Indians of Puget Sound. *University of Washington Publications in Anthropology* 4(1):1–84.

Haines, Aubrey L. 1962. *Mountain Fever: Historic Conquests of Rainier.* Portland: Oregon Historical Society.

Hall, E. Raymond. 1981. *The Mammals of North America.* 2nd edition. 2 vols. New York: John Wiley and Sons.

Hall, E. Raymond, and Keith R. Kelson. 1959. *The Mammals of North America.* 2 vols. New York: Ronald Press.

Hall, E. S. 1932. A Historical Resume of Exploration and Survey—Mammal Types and Their Collectors in the State of Washington. *Murrelet* 13:63–91.

Hallock, Charles. 1879. *The Sportsman's Gazetteer and General Guide.* New York: Forest and Stream Publishing Company.

Halvorson, William L., and Gary E. Davis (eds.). 1996. *Science and*

Ecosystem Management in the National Parks. Tucson: University of Arizona Press.

Harington, C. R. 1971. A Pleistocene Mountain Goat from British Columbia and Comments on the Dispersal History of *Oreamnos. Canadian Journal of Earth Sciences* 8:1081–93.

Heaton, Timothy H. 1994. Variation in Fossil and Modern *Ursus* from Alaska (Abstract). *Journal of Vertebrate Paleontology* (Supplement) 14:28A.

——. 1995. Colonization of Southeast Alaska by *Ursus arctos* Prior to the Peak of Wisconsin Glaciation (Abstract). *Journal of Vertebrate Paleontology* (Supplement) 15:134A.

——. 1996. The Late Wisconsin Vertebrate Fauna of On Your Knees Cave, Northern Prince of Wales Island, Southeast Alaska (Abstract). *Journal of Vertebrate Paleontology* (Supplement) 16:40A.

——. 1997. The Preliminary Late Wisconsin Mammalian Biochronology of Prince of Wales Island, Southeastern Alaska (Abstract). *Journal of Vertebrate Paleontology* (Supplement) 17:52A.

Heaton, Timothy H., and Frederick Grady. 1992a. Preliminary Report on the Fossil Bears of El Capitan Cave, Prince of Wales Island, Alaska. *Current Research in the Pleistocene* 9:97–99.

——. 1992b. Two Species of Bear Found in Late Pleistocene/Early Holocene Den in El Capitan Cave, Prince of Wales Island, Southern Alaskan Coast (Abstract). *Journal of Vertebrate Paleontology* (Supplement) 12:32A.

——. 1993. Fossil Grizzly Bears (*Ursus arctos*) from Prince of Wales Island, Alaska, Offer New Insights into Animal Dispersal, Interspecific Competition, and Age of Deglaciation. *Current Research in the Pleistocene* 10:98–100.

Heaton, Timothy H., Sandra L. Talbott, and Gerald F. Shields. 1996. An Ice Age Refugium for Large Mammals in the Alexander Archipelago, Southeastern Alaska. *Quaternary Research* 46:186–92.

Hebert, Daryl, and I. McTaggart Cowan. 1971. Natural Salt Licks as a Part of the Ecology of the Mountain Goat. *Canadian Journal of Zoology* 49:605–10.

Hebert, D. M., and W. G. Turnbull. 1977. A Description of Southern Interior and Coastal Mountain Goat Ecotypes in British Columbia. In *Proceedings of the First International Mountain Goat Symposium,* edited by William Samuel and W. G. Macgregor, pp.

126–46. Victoria, B.C.: Province of British Columbia, Ministry of Recreation and Conservation, Fish and Wildlife Branch.

Henderson, Jan A. 1984. Goat Salting Impact Study. Unpublished report on file, Olympic National Park, Port Angeles.

Hess, Karl, Jr. 1993. *Rocky Times in Rocky Mountain National Park.* Niwot: University of Colorado Press.

Hester, F. Eugene. 1991. The U.S. National Park Service Experience with Exotic Species. *Natural Areas Journal* 11:127–28.

Heusser, Calvin J. 1983. Vegetational History of the Northwestern United States, including Alaska. In *Late Quaternary Environments of the United States, Vol. 1, The Late Pleistocene*, edited by Stephen C. Porter, pp. 239–58. Minneapolis: University of Minnesota Press.

Higgs, Eric S. 1997. What Is Good Ecological Restoration? *Conservation Biology* 11:338–48.

Hill, Edward P. 1982. Beaver: *Castor canadensis.* In *Wild Mammals of North America: Biology, Management, Economics*, edited by Joseph A. Chapman and George A. Feldhamer, pp. 256–81. Baltimore: Johns Hopkins University Press.

Hoerr, Winfried. 1993. The Concept of Naturalness in Environmental Discourse. *Natural Areas Journal* 13:29–32.

Hoffman, Roger A., and R. Gerald Wright. 1987. A Comparison of Population Control Techniques for Mountain Goats in Olympic National Park, Using Field Sterilization Procedures and Computer Modeling. Unpublished report on file, Olympic National Park, Port Angeles.

——. 1990. Fertility Control in a Non-native Population of Mountain Goats. *Northwest Science* 64:1–6.

Hoffmann, R. S., and C. F. Nadler. 1968. Chromosomes and Systematics of Some North American Species of the Genus *Marmota* (Rodentia: Sciuridae). *Experientia* 24:740–42.

Hoffmann, Robert S., James W. Kloeppl, and Charles F. Nadler. 1979. The Relationships of the Amphiberingian Marmots (Mammalia: Sciuridae). *Occasional Papers of the University of Kansas Museum of Natural History* (Lawrence) 83:1–56.

Hornaday, W. T. 1904. *The American Natural History* (1922 edition). New York: Charles Scribner's Sons.

Houston, Douglas B. 1971. Ecosystems of National Parks. *Science* 172:648–51.

——. 1995. Response to "Inaccurate Data and the Olympic National Park Mountain Goat Controversy." *Northwest Science* 69:239–40.

Houston, D. B., R. A. Hoffman, R. W. Olson, and B. B. Moorhead. 1994. Mountain Goat Management in Olympic National Park. In *Mountain Goats in Olympic National Park: Biology and Management of an Introduced Species,* by Douglas B. Houston, Edward G. Schreiner, and Bruce B. Moorhead, pp. 190–208. Denver: USDI–NPS Scientific Monograph NPS/NROLYM/NRSM-94/25.

Houston, D. B., B. B. Moorhead, and R. W. Olson. 1986. An Aerial Census of Mountain Goats in the Olympic Range, Washington. *Northwest Science* 60:131–36.

——. 1991. Mountain Goat Population Trends in the Olympic Mountain Range, Washington. *Northwest Science* 65:212–16.

Houston, D. B., R. W. Olson, R. A. Hoffman, and B. B. Moorhead. 1994 Mountain Goat Census in the Olympic Mountain Range, 1994. Unpublished report on file, Olympic National Park, Port Angeles.

Houston, Douglas B., and Edward G. Schreiner. 1993. Biogeography of the Olympic Peninsula. Unpublished manuscript on file, Olympic National Park, Port Angeles.

—— (and N. M. Buckingham). 1994a. Biogeography of the Olympic Peninsula. In *Mountain Goats in Olympic National Park: Biology and Management of an Introduced Species,* by Douglas B. Houston, Edward G. Schreiner, and Bruce B. Moorhead, pp. 28–46. Denver: USDI–NPS Scientific Monograph NPS/NROLYM/NRSM-94/25.

——. 1994b. Introduction. In *Mountain Goats in Olympic National Park: Biology and Management of an Introduced Species,* by Douglas B. Houston, Edward G. Schreiner, and Bruce B. Moorhead, pp. 5–12. Denver: USDI–NPS Scientific Monograph NPS/NROLYM/NRSM-94/25.

——. 1995. Alien Species in National Parks: Drawing Lines in Space and Time. *Conservation Biology* 9:204–9.

Houston, Douglas B., Edward G. Schreiner, and Bruce B. Moorhead. 1994. *Mountain Goats in Olympic National Park: Biology and Management of an Introduced Species.* Denver: USDI–NPS Scientific Monograph NPS/NROLYM/NRSM-94/25.

Houston, D. B., E. G. Schreiner, B. B. Moorhead, and K. A. Krueger. 1990. Elk in Olympic National Park: Will They Persist over Time? *Natural Areas Journal* 10:6–11.

Houston, Douglas B., Edward G. Schreiner, Bruce B. Moorhead, and Richard W. Olson. 1991. Mountain Goat Management in Olympic National Park: A Progress Report. *Natural Areas Journal* 11:87–92.

Houston, D. B., and V. Stevens. 1988. Resource Limitation in Mountain Goats: A Test by Experimental Cropping. *Canadian Journal of Zoology* 66:228–38.

Howell, Benita J. 1994. Linking Cultural and Natural Conservation in National Park Service Policies and Programs. In *Conserving Culture: A New Discourse on Heritage*, edited by Mary Hufford, pp. 122–37. Urbana: University of Illinois Press.

Hubbard, R. P. 1967. Preliminary Report on the Archaeological Survey of the Owyhee Uplands in Idaho and Oregon, 1965–67. Unpublished report on file, Department of Anthropology, Oregon State University, Corvallis.

Huelsbeck, David R. 1983. Mammals and Fish in the Subsistence Economy of Ozette. Ph.D. dissertation, Washington State University, Pullman.

——. 1994. Mammals and Fish in the Subsistence Economy of Ozette. In *Ozette Archaeological Project Research Reports, Volume II: Fauna*, edited by Stephan R. Samuels, pp. 17–91. Reports of Investigations 66. Department of Anthropology, Washington State University, Pullman, and National Park Service, Pacific Northwestern Regional Office, Seattle.

Hult, Ruby El. 1954. *The Untamed Olympics*. Portland, Ore.: Binfords and Mort.

Huntington, Dwight W. 1904. *Our Big Game*. New York: Charles Scribner's Sons.

Hutchins, Michael. 1984. The Mother-Offspring Relationships in Mountain Goats (*Oreamnos americanus*). Ph.D. dissertation, University of Washington, Seattle.

——. 1995. Olympic Mountain Goat Controversy Continues. *Conservation Biology* 9:1324–26.

Hutchins, Michael, and Valerius Geist. 1987. Behavioural Considerations in the Management of Mountain-Dwelling Ungulates. *Mountain Research and Development* 7:135–44.

Hutchins, Michael, and Victoria Stevens. 1981. Olympic Mountain Goats. *Natural History* 90(1):58–69.

Hyland, D. C., J. M. Tersak, J. M. Adovasio, and M. I. Siegal. 1990.

Identification of the Species of Origin of Residual Blood on Lithic Material. *American Antiquity* 55:104–12.

In the Olympic Mountains. 1890. *Oregonian*, December 31.

Introduction of Mountain Goats in the Olympic Mountains. 1925. *Murrelet* 6(1):10.

Irby, M. Lanette, and Alex F. Chappell. 1994. Review of the Historical Literature regarding the Distribution of the Rocky Mountain Goat (*Oreamnos americanus*). In *Proceedings of the Ninth Biennial Symposium of the Northern Wild Sheep and Goat Council*, edited by Margo Pybus, pp. 75–80. Northern Wild Sheep and Goat Council.

Johnson, Donald R. 1961. Additional Record of Mountain Goat in Northeastern Washington. *Journal of Mammalogy* 42:280.

———. 1970. Further Evidence of Mountain Goat in the Southern Cascades. *Murrelet* 51(1):13.

Johnson, Kajsa N., Rebecca L. Johnson, Daniel K. Edwards, and Christopher A. Wheaton. 1993. Public Participation in Wildlife Management: Opinions from Public Meetings and Random Surveys. *Wildlife Society Bulletin* 21:218–25.

Johnson, Murray L. 1968. Application of Blood Protein Electrophoretic Studies to Problems in Mammalian Taxonomy. *Systematic Zoology* 17:23–30.

Johnson, Murray L., and Seth B. Benson. 1960. Relationship of the Pocket Gophers of the *Thomomys mazama-talpoides* Complex in the Pacific Northwest. *Murrelet* 41(2):17–22.

Johnson, Murray [L.], and Sherry Johnson. 1952. Check List of Mammals of the Olympic Peninsula. *Murrelet* 33:32–37.

Johnson, Murray L., and Burton O. Ostenson. 1959. Comments on the Nomenclature of Some Mammals of the Pacific Northwest. *Journal of Mammalogy* 40:571–77.

Johnson, Murray L., and Terry L. Yates. 1980. A New Townsend's Mole (Insectivora: Talpidae) from the State of Washington. *Occasional Papers of the Museum of Texas Tech University* (Lubbock) 63:1–6.

Johnson, Rolf L. 1977. Distribution, Abundance and Management Status of Mountain Goats in North America. In *Proceedings of the First International Mountain Goat Symposium*, edited by William Samuel and W. G. Macgregor, pp. 1–7. Victoria, B.C.: Province of British Columbia, Ministry of Recreation and Conservation, Fish and Wildlife Branch.

——. 1983. *Mountain Goats and Mountain Sheep of Washington.* Biological Bulletin No. 18. Olympia: Washington State Game Department.

——. 1986. Mountain Goat Management in Washington. *Proceedings of the Biennial Symposium of the Northern Wild Sheep and Goat Council* 5:60–62.

Johnson, Rolf, and Bruce Moorhead. 1982. Trapping and Transplanting Mountain Goats in Washington State. *Proceedings of the Biennial Symposium of the Northern Wild Sheep and Goat Council* 3:106–14.

Johst, Karin, and Roland Brandl. 1997. Body Size and Extinction Risk in a Stochastic Environment. *Oikos* 78:612–17.

Jokerst, B. James. 1960. Distribution and Population of the Mountain Goat in Olympic National Park. Unpublished report on file, Olympic National Park, Port Angeles.

Kay, Charles E. 1990. Yellowstone's Northern Elk Herd: A Critical Evaluation of the "Natural Regulation" Paradigm. Ph.D. dissertation, Utah State University, Logan.

——. 1994. Aboriginal Overkill: The Role of Native Americans in Structuring Western Ecosystems. *Human Nature* 5:359–98.

Kaye, Thomas N. 1989. Autecology, Reproductive Ecology, and Demography of *Astragalus australis* var. *olympicus* (Fabaceae). M.S. thesis, Oregon State University, Corvallis.

King, Arden R. 1950. *Cattle Point: A Stratified Site in the Southern Northwest Coast Region.* Society for American Archaeology Memoir No. 7.

Kirk, Ruth and Richard D. Daugherty. 1978. *Exploring Washington Archaeology.* Seattle: University of Washington Press.

Klein, David R. 1965. Postglacial Distribution Patterns of Mammals in the Southern Coastal Regions of Alaska. *Arctic* 18:7–20.

Kooyman, Brian, and Margaret E. Newman. 1992. Verifying the Reliability of Blood Residue Analysis on Archaeological Tools. *Journal of Archaeological Science* 19:265–69.

Korn, H. 1994. Genetic, Demographic, Spatial, Environmental and Catastrophic Effects on the Survival Probability of Small Populations of Mammals. In *Minimum Animal Populations,* edited by Hermann Remmert, pp. 33–49. Berlin: Springer-Verlag.

Krebs, Charles J. 1988. *The Message of Ecology.* New York: Harper and Row.

Krug, J. A. 1946. Dedication of the Olympic National Park. *Journal of Forestry* 44:549–51.

Labisky, Ronald F., Dale Stanbury, and Stephen C. Smith. 1986. Fish and Wildlife: The Forgotten Resource in National Policy. *Transactions of the Fifty-first North American Wildlife and Natural Resources Conference*, pp. 538–42. Washington, D.C.: Wildlife Society.

Lacava, Jerry, and Jeff Hughes. 1984. Determining Minimum Viable Population Levels. *Wildlife Society Bulletin* 12:370–76.

Lacy, Robert C. 1997. Importance of Genetic Variation to the Viability of Mammalian Populations. *Journal of Mammalogy* 78:320–35.

Lande, Russell. 1988. Genetics and Demography in Biological Conservation. *Science* 241:1455–60.

——. 1993. Risks of Population Extinction from Demographic and Environmental Stochasticity and Random Catastrophes. *American Naturalist* 142:911–27.

Langemann, Elisabeth G. 1987. Zooarchaeology of the Lillooet Region, British Columbia. M.A. thesis, Simon Fraser University, Burnaby, British Columbia.

Larrison, Earl J., and Klaus G. Sonnenberg. 1968. *Washington Birds: Their Location and Identification.* Seattle: Seattle Audobon Society.

Laundré, John W. 1991. Mountain Goats in Yellowstone: The Horns of a Dilemma? *Park Science* 11(3):8–9.

——. 1994. Resource Overlap between Mountain Goats and Bighorn Sheep. *Great Basin Naturalist* 54:114–21.

Lee, John A. 1913. The Mountaineers' Outing in the Olympics. *Mazama* 4(2):70–75.

Lee, Kai N. 1993. *Compass and Gyroscope: Integrating Science and Politics for the Environment.* Washington, D.C.: Island Press.

Lemons, John. 1987. United States' National Park Management: Values, Policy, and Possible Hints for Others. *Environmental Conservation* 14:329–40.

Leonhardy, Frank C., and R. Wayne Thompson. 1991. *Archaeological Investigations at 35WA286 and 35WA288, Hells Canyon National Recreation Area, Wallowa County, Oregon.* Alfred M. Bowers Laboratory of Anthropology Letter Report 91-11. Moscow: University of Idaho.

Leopold, A. S., S. A. Cain, C. M. Cottam, I. N. Gabrielson, and T. L. Kimball. 1963. Wildlife Management in the National Parks. *Transactions of the Twenty-eighth North American Wildlife and Nat-*

ural Resources Conference, pp. 28–45. Washington, D.C.: Wildlife Society.

Livingston, Stephanie D. 1987. Prehistoric Biogeography of White-Tailed Deer in Washington and Oregon. *Journal of Wildlife Management* 51:649–54.

Lomolino, Mark V. 1993. Winter Filtering, Immigrant Selection and Species Composition of Insular Mammals of Lake Huron. *Ecography* 16:24–30.

——. 1994. Species Richness of Mammals Inhabiting Nearshore Archipelagoes: Area, Isolation, and Immigration Filters. *Journal of Mammalogy* 75:39–49.

——. 1996. Investigating Causality of Nestedness of Insular Communities: Selective Immigrations or Extinctions? *Journal of Biogeography* 23:699–703.

Lomolino, Mark V., James H. Brown, and Russell Davis. 1989. Island Biogeography of Montane Forest Mammals in the American Southwest. *Ecology* 70:180–94.

Loy, Thomas H. 1983. Prehistoric Blood Residues: Detection on Tool Surfaces and Identification of Species of Origin. *Science* 220: 1269–71.

Lundelius, Ernest L., Jr., Russell W. Graham, Elaine Anderson, John Guilday, J. Alan Holman, David W. Steadman, and S. David Webb. 1983. Terrestrial Vertebrate Faunas. In *Late Quaternary Environments of the United States, Vol. 1, The Late Pleistocene*, edited by Stephen C. Porter, pp. 311–53. Minneapolis: University of Minnesota Press.

Luoma, Jon R. 1993. Gunning for Goats. *Wildlife Conservation* 96(6):30–41.

Lyman, R. Lee. 1986. On the Holocene History of *Ursus* in Eastern Washington. *Northwest Science* 60:67–72.

——. 1988a. Significance for Wildlife Management of the Late Quaternary Biogeography of Mountain Goats (*Oreamnos americanus*) in the Pacific Northwest, U.S.A. *Arctic and Alpine Research* 20:13–23.

——. 1988b. Zoogeography of Oregon Coast Marine Mammals: The Last 3000 Years. *Marine Mammal Science* 4:247–64.

——. 1989. Natives or Aliens? *Pacific Discovery* 42(3):3.

——. 1991. The Holocene History of the Red Fox (*Vulpes vulpes*) in Eastern Washington. *Northwest Science* 65:22–26.

——. 1994. The Olympic Mountain Goat Controversy: A Different Perspective. *Conservation Biology* 8:898–901.

——. 1995a. Inaccurate Data and the Olympic Mountain Goat Controversy. *Northwest Science* 69:234–38.

——. 1995b. Determining When Rare (Zoo) Archaeological Phenomena Are Truly Absent. *Journal of Archaeological Method and Theory* 2:369–424.

Lyman, R. Lee, and Stephanie D. Livingston. 1983. Late Quaternary Mammalian Zoogeography of Eastern Washington. *Quaternary Research* 20:360–73.

MacArthur, Robert H. 1972. *Geographical Ecology: Patterns in the Distribution of Species.* Princeton: Princeton University Press.

MacArthur, Robert H., and Edward O. Wilson. 1967. *The Theory of Island Biogeography.* Monographs in Population Biology 1. Princeton: Princeton University Press.

MacDonald, Stephen O., and Joseph A. Cook. 1996. The Land Mammal Fauna of Southeast Alaska. *Canadian Field-Naturalist* 110:571–98.

Machlis, G. E., S. Tuler, and R. Kasperson. 1990. A Social Risk Analysis of the Olympic National Park Mountain Goat Removal Project. University of Idaho, Cooperative Park Studies Unit, Unpublished report on file, Olympic National Park, Port Angeles (NPS Reference #D-173).

Macnab, John. 1983. Wildlife Management as Scientific Experimentation. *Wildlife Society Bulletin* 11:397–401.

——. 1985. Carrying Capacity and Related Slippery Shibboleths. *Wildlife Society Bulletin* 13:403–10.

Macy, Preston P. 1934. Some Notes on the Animal Life of Mount Rainier National Park, Washington. *Murrelet* 15:46–48.

Maddox, John. 1989. Where Next with Peer-Review? *Nature* 339:11.

Majors, Harry M. (ed.). 1981a. The Press Party Expedition in the Olympic Mountains, 1889–1890. *Northwest Discovery* 2(2):71–134.

——. 1981b. The Press Party Expedition in the Olympic Mountains, 1889–1890, continued. *Northwest Discovery* 2(3):138–98.

——. 1981c. The Press Party Expedition in the Olympic Mountains: Devils Backbone, February 1890. *Northwest Discovery* 2(7):465–74.

——. 1981d. The Press Party Expedition in the Olympic Mountains: The Limits of Settlement, February 1890. *Northwest Discovery* 2(5):314–21.

——. 1984a. The Press Party Expedition in the Olympic Mountains,

1889–1890, Part 3—Winter on the Elwha River. *Northwest Discovery* 5(23):202–86.

——. 1984b. The Press Party Expedition in the Olympic Mountains, 1889–1890, Part 4—The Center of the Olympic Mountains. *Northwest Discovery* 5(24):290–364.

——. 1984c. The Press Party Expedition in the Olympic Mountains, 1889–1890, Part 5—Down the Quinault River. *Northwest Discovery* 5(25):368–448.

Manville, Richard H., and Stanley P. Young. 1965. *Distribution of Alaskan Mammals.* Washington, D.C.: USDI, Bureau of Sports Fisheries and Wildlife Circular 221.

Martin, Paul S. 1970. Pleistocene Niches for Alien Animals. *BioScience* 20:218–21.

Maser, Chris. 1990. On the "Naturalness" of Natural Areas: A Perspective for the Future. *Natural Areas Journal* 10:129–33.

Mathewes, Rolf W. 1993. Evidence for Younger Dryas-Age Cooling on the North Pacific Coast of America. *Quaternary Science Reviews* 12:321–31.

Mathias, F. W. 1928. The Olympic Mountains. *Mazama* 10(12):33–43.

Matter, William J., and R. William Mannan. 1989. More on Gaining Reliable Knowledge: A Comment. *Journal of Wildlife Management* 53:1172–76.

Matthews, Olen Paul. 1986. Who Owns Wildlife? *Wildlife Society Bulletin* 14:459–65.

Matthews, Patrick E., and Victor L. Coggins. 1994. Status and History of Mountain Goats in Oregon. In *Proceedings of the Ninth Biennial Northern Wild Sheep and Goat Council,* edited by Margo Pybus, pp. 69–74. Northern Wild Sheep and Goat Council.

Mattson, David J. 1996. Ethics and Science in Natural Resource Agencies. *BioScience* 46:767–71.

Mayr, Ernst. 1993. Fifty Years of Progress in Research on Species and Speciation. *Proceedings of the California Academy of Sciences* 48:131–40.

McCain, Garvin, and Erwin M. Segal. 1969. *The Game of Science.* Belmont, Calif.: Wadsworth Publishing.

McDonald, Kelly A., and James H. Brown. 1992. Using Montane Mammals to Model Extinctions Due to Global Change. *Conservation Biology* 6:409–15.

McLeod, Rebecca Lynn Malone. 1984. An American Wilderness: The Origins of Olympic National Park. M.A. thesis, University of California, Santa Barbara.

McNulty, Tim. 1996. *A Natural History Guide: Olympic National Park.* Boston: Houghton Mifflin.

Mead, Jim I. 1983. Harrington's Extinct Mountain Goat (*Oreamnos harringtoni*) and Its Environment in the Grand Canyon, Arizona. Ph.D. dissertation, University of Arizona, Tucson.

———. 1987. Quaternary Records of Pika, *Ochotona*, in North America. *Boreas* 16:165–71.

Mead, Jim I., Larry D. Agenbroad, Arthur M. Phillips III, and Larry T. Middleton. 1987. Extinct Mountain Goat (*Oreamnos harringtoni*) in Southeastern Utah. *Quaternary Research* 27:323–31.

Mead, Jim I., Christopher J. Bell, and Lyndon K. Murray. 1992. *Mictomys borealis* (Northern Bog Lemming) and the Wisconsin Paleoecology of the East-Central Great Basin. *Quaternary Research* 37:229–38.

Mead, Jim I., and Mark C. Lawler. 1994. Skull, Mandible, and Metapodials of the Extinct Harrington's Mountain Goat (*Oreamnos harringtoni*). *Journal of Vertebrate Paleontology* 14:562–76.

Mead, Jim I., Paul S. Martin, Robert C. Euler, Austin Long, A. J. T. Jull, Laurence J. Toolin, Douglas J. Donahue, and T. W. Linick. 1986. Extinction of Harrington's Mountain Goat. *Proceedings of the National Academy of Science* 83:836–39.

Mead, Jim I., Mary K. O'Rourke, and Theresea M. Foppe. 1986. Dung and Diet of the Extinct Harrington's Mountain Goat (*Oreamnos harringtoni*). *Journal of Mammalogy* 67:284–93.

Meltzer, David J. 1993. *Search for the First Americans.* Washington, D.C.: Smithsonian Institution.

Merriam, C. Hart. 1895. The Earliest Available Name for the Mountain Goat. *Science* 1:19.

———. 1897a. Suggestions for a New Method of Discriminating between Species and Subspecies. *Science* 5:753–58.

———. 1897b. Synonymy of the Mountain Goat (in a letter to E. M. Wilbur, Secretary). *Mazama* 1(2):288–89.

———. 1902. The Mountain Goat. *Country Life in America* (n.s.) 2(1):32.

———. 1919. Criteria for the Recognition of Species and Genera. *Journal of Mammalogy* 1:6–9.

Merritt, Jim. 1995. Look Back: The Early Years 1895–1914. *Field and Stream* 99(9):42–50.

Michalovic, Joanne. 1984. Summer Distribution and Abundance of Mountain Goats in Mt. Rainier National Park, Washington. In

Proceedings of the Fourth Biennial Northern Wild Sheep and Goat Council, edited by Manfred Hoefs, pp. 137–66. Whitehorse: Yukon Wildlife Branch.

Mierendorf, Robert R., and David J. Harry. 1993. Results of Subsurface Archaeological Survey in a Pleistocene Terrace in North Cascades National Park Service Complex. *Archaeology in Washington* 5:39–51.

Mills, L. Scott, and Peter E. Smouse. 1994. Demographic Consequences of Inbreeding in Remnant Populations. *American Naturalist* 144:412–31.

Moodie, G. Eric E., and Thomas E. Reimchen. 1976. Glacial Refugia, Endemism, and Stickleback Populations of the Queen Charlotte Islands, British Columbia. *Canadian Field-Naturalist* 90:471–74.

Moorhead, Bruce B. 1977. The Enigmatic Mountain Goat. *Pacific Search* 11(8):2–3.

——. 1988a. Non-native Mountain Goat Management Undertaken at Olympic National Park. *Park Science* 9(3):10–11.

——. 1988b. Review of Paper by R. Lee Lyman about the Likelihood of Prehistoric Mountain Goats in the Park and Rethinking NPS Wildlife Management Policies. Unpublished interdepartmental Olympic National Park memorandum.

Moorhead, Bruce B., and Edward S. Schreiner. 1979. Management Studies of Human Impact at Backcountry Campsites in Olympic National Park, Washington. In *Proceedings of the First Conference on Scientific Research in the National Parks*, vol. 2, edited by Robert M. Linn, pp. 1273–78. Transactions and Proceedings Series No. 5. Washington, D.C.: National Park Service.

Moorhead, Bruce B., and Victoria Stevens. 1982. Introduction and Dispersal of Mountain Goats in Olympic National Park. In *Ecological Research in National Parks of the Pacific Northwest*, edited by Edward E. Starkey, Jerry F. Franklin, and Jean W. Matthews, pp. 46–50. Corvallis: Forest Research Laboratory, Oregon State University.

Morgenroth, Chris. 1991. *Footprints in the Olympics: An Autobiography*. Edited by Katherine Morgenroth Flaherty. Fairfield, Wash.: Ye Galleon Press.

Mountain Goats May Be Introduced. 1919. *Port Angeles Evening News*, Oct. 16.

Mount Rainier and Olympic National Parks Magazine. 1994. 3rd ed. San Francisco: American Parks Network.

Murphy, Dennis D., and Barry D. Noon. 1991. Coping with Uncertainty in Wildlife Biology. *Journal of Wildlife Management* 55:773–82.

Mustoe, George E., and Cynthia A. Carlstad. 1995. A Late Pleistocene Brown Bear (*Ursus arctos*) from Northwest Washington. *Northwest Science* 69:106–13.

Nagorsen, David. 1990. *The Mammals of British Columbia: Taxonomic Catalogue.* Memoir 4. Victoria: Royal British Columbia Museum.

Nagorsen, David W., Grant Keddie, and Tanya Luszcz. 1996. *Vancouver Island Marmot Bones from Subalpine Caves: Archaeological and Biological Significance.* Occasional Paper No. 4. Victoria, B.C.: B.C. Parks.

National Park Service. 1993. *Science and the National Parks II: Adapting to Change.* Washington, D.C.: U.S. Government Printing Office.

National Research Council. 1992. *Science and the National Parks.* Washington, D.C.: National Academy of Sciences.

Naylor, Kirk S., James M. Peek, and Peter Zager. 1990. Perspectives on Management of Isolated Mountain Goat Populations with High Nonconsumptive Values: The Pend Oreille Lake, Idaho, Case Study. *Proceedings of the Biennial Symposium of the Northern Wild Sheep and Goat Council* 7:282–89.

Newman, Coleman C. 1954. Special Report on Mountain Goat. Unpublished report on file, Olympic National Park, Port Angeles.

Newmark, William D. 1986a. Mammalian Richness, Colonization, and Extinction in Western North American National Parks. Ph.D. dissertation, University of Michigan, Ann Arbor.

——. 1986b. Species-Area Relationship and Its Determinants for Mammals in Western North American National Parks. In *Island Biogeography of Mammals,* edited by L. R. Heaney and Bruce D. Patterson, pp. 83–98. Biological Journal of the Linnean Society 28.

——. 1987. A Land-Bridge Island Perspective on Mammalian Extinctions in Western North American Parks. *Nature* 325:430–32.

——. 1995. Extinction of Mammal Populations in Western North American National Parks. *Conservation Biology* 9:512–26.

Noss, Reed F. 1996. Conservation Biology, Values, and Advocacy. *Conservation Biology* 10:904–20.

Nudds, Thomas D., and Michael L. Morrison. 1991. Ten Years after "Reliable Knowledge": Are We Gaining? *Journal of Wildlife Management* 55:757–60.

Nunney, Leonard, and Diane R. Elam. 1994. Estimating the Effective Population Size of Conserved Populations. *Conservation Biology*

8:175–84.

O'Brien, Mary H. 1993. The Professional Biologist: Being Scientists Means Taking Sides. *BioScience* 43:706–8.

O'Brien, Stephen J., and Ernst Mayr. 1991. Bureaucratic Mischief: Recognizing Endangered Species and Subspecies. *Science* 251:1187–88.

Olmsted, Ingrid C. 1977. Interrelationships of Introduced Mountain Goats and Subalpine Habitat in Olympic National Park. In *Proceedings of the First International Mountain Goat Symposium*, edited by William Samuel and W. G. Macgregor, p. 200. Victoria, B.C.: Province of British Columbia, Ministry of Recreation and Conservation, Fish and Wildlife Branch.

——. 1979. Alpine and Subalpine Vegetation under the Influence of Non-native Mountain Goats, Olympic National Park. In *Proceedings of the First Conference on Scientific Research in the National Parks*, vol. 2, edited by Robert M. Linn, pp. 1143–48. Transactions and Proceedings Series No. 5. Washington, D.C.: National Park Service.

Olson, Storrs L. 1989. Extinction on Islands: Man as a Catastrophe. In *Conservation for the Twenty-first Century*, edited by David Western and Mary C. Pearl, pp. 50–53. New York: Oxford University Press.

Olympic National Park. 1981. An Environmental Assessment on the Management of Introduced Mountain Goats in Olympic National Park. Unpublished report on file, Olympic National Park, Port Angeles.

——. 1982. Mountain Goat Ecology and Management Investigations, Olympic National Park, 1981. Unpublished report on file, Olympic National Park, Port Angeles.

——. 1983. Mountain Goat Ecology and Management Investigations, Olympic National Park, 1982. Unpublished report on file, Olympic National Park, Port Angeles.

——. 1984. Mountain Goat Ecology and Management Investigations, Olympic National Park, 1983. Unpublished report on file, Olympic National Park, Port Angeles.

——. 1987a. Environmental Assessment: Mountain Goat Management in Olympic National Park. Unpublished report on file, Olympic National Park, Port Angeles.

——. 1987b. The Management of Mountain Goats in Olympic National Park: Some Questions and Some Answers. Circulated pamphlet, Olympic National Park, Port Angeles.

——. 1995. *Goats in Olympic National Park: Draft Environmental Impact Statement for Mountain Goat Management.* Port Angeles, Wash.: USDI–NPS, Olympic National Park.

Olympic National Park Created. 1938. *American Forests* 44(8):376.

O'Neil, Joseph. 1896. *O'Neil Expedition.* 54th Congress, 1st Session, Document No. 59.

Osgood, Wilfred H. 1943. Clinton Hart Merriam—1855–1942 (obituary). *Journal of Mammalogy* 24:421–36.

——. 1947. Biographical Memoir of Clinton Hart Merriam, 1855–1942. *National Academy of Sciences of the United States of America Biographical Memoirs* 24(1):1–57.

Palmer, T. S. 1954. In Memoriam: Clinton Hart Merriam. *Auk* 71:130–36.

Patterson, Bruce D. 1984. Mammalian Extinction and Biogeography in the Southern Rocky Mountains. In *Extinctions,* edited by Matthew H. Nitecki, pp. 247–93. Chicago: University of Chicago Press.

——. 1987. The Principle of Nested Subsets and Its Implications for Biological Conservation. *Conservation Biology* 1:323–34.

——. 1990. On the Temporal Development of Nested Subset Patterns of Species Composition. *Oikos* 59:330–42.

Patterson, Bruce D., and Wirt Atmar. 1986. Nested Subsets and the Structure of Insular Mammalian Faunas and Archipelagos. In *Island Biogeography of Mammals,* edited by L. R. Heaney and Bruce D. Patterson, pp. 65–82. Biological Journal of the Linnean Society 28.

Peek, James M., Dale G. Miquelle, and R. Gerald Wright. 1987. Are Bison Exotic in the Wrangell–St. Elias National Park and Preserve? *Environmental Management* 11:149–53.

Peters, William Harrison. 1936. In the Forests of the Olympics. *American Forests* 42:170–76.

Petersen, K., P. J. Mehringer, Jr., and Carl E. Gustafson. 1983. Late-Glacial Vegetation and Climate at the Manis Mastodon Site, Olympic Peninsula, Washington. *Quaternary Research* 20:215–31.

Peterson, David, and Andrea Woodward. 1993. Face the Facts: Olympic Goats Must Go. *Seattle Times,* October 22.

Peterson, David L., David G. Silsbee, and Daniel L. Schmoldt. 1994. A Case Study of Resources Management Planning with Multiple Objectives and Projects. *Environmental Management* 18:729–42.

Pfitsch, W. A. 1981. The Effect of Mountain Goats on the Subalpine Plant Communities of Klahhane Ridge, Olympic National Park,

Washington. M.S. thesis, University of Washington, Seattle.

Pfitsch, W. A., and L. C. Bliss. 1985. Seasonal Forage Availability and Potential Vegetation Limitations to a Mountain Goat Population, Olympic National Park. *American Midland Naturalist* 113:109–21.

Pfitsch, W. A., R. S. Reid, J. Harter, D. K. Pike, and L. C. Bliss (eds.). 1983. Effects of Mountain Goats on Soils, Plant Communities and Selected Species in Olympic National Park. Unpublished report on file, Olympic National Park, Port Angeles.

Phillips, Mike. 1996. Don't Let Them Get Our Goats. *Bremerton Sun*, Dec. 16, 1996.

Pielou, E. C. 1991. *After the Ice Age: The Return of Life to Glaciated North America.* Chicago: University of Chicago Press.

Pike, D. K. 1981. Effects of Mountain Goats on Three Plant Species Unique to the Olympic Mountains, Washington. M.S. thesis, University of Washington, Seattle.

Pimm, Stuart L., H. Lee Jones, and Jared Diamond. 1988. On the Risk of Extinction. *American Naturalist* 132:757–85.

Porter, Stephen C., Kenneth L. Pierce, and Thomas D. Hamilton. 1983. Late Wisconsin Mountain Glaciation in the Western United States. In *Late Quaternary Environments of the United States, Vol. 1, The Late Pleistocene*, edited by Stephen C. Porter, pp. 71–111. Minneapolis: University of Minnesota Press.

Rainbolt, Raymond E., and Bruce E. Coblentz. 1997. A Different Perspective on Eradication of Vertebrate Pests. *Wildlife Society Bulletin* 25:189–91.

Randolph, J. E., and M. Dahlstrom. 1977. *Archaeological Test Excavations at Bernard Creek Rockshelter.* Anthropological Research Manuscript Series No. 42. Moscow: University of Idaho.

Raup, David M., and Rex E. Crick. 1979. Measurement of Faunal Similarity in Paleontology. *Journal of Paleontology* 53:1213–27.

Reagan, Albert B. 1908a. Animals of the Olympic Peninsula, Washington. *Proceedings of the Indiana Academy of Sciences for 1908*:193–99.

——. 1908b. The Olympic Region. *Alaska-Yukon Magazine* 6(1):18–23.

——. 1909. Notes on the Olympic Peninsula, Washington. *Transactions of the Kansas Academy of Sciences* 22:131–238.

——. 1917. Archaeological Notes on Western Washington and Adjacent British Columbia. *Proceedings of the California Academy of Sciences*, fourth series 7(1):1–31.

Reid, R. S. 1983. The Effect of Mountain Goats on Plant Establishment and Seedling Survival, Olympic National Park. M.S. thesis, University of Washington, Seattle.

Reiger, John F. 1992. Wildlife, Conservation, and the First Forest Reserve. In *Origins of the National Forests: A Centennial Symposium*, edited by Harold K. Stern, pp. 106–21. Durham, N.C.: Forest History Society.

Rensberger, John M., and Anthony D. Barnosky. 1993. Short-term Fluctuations in Small Mammals of the Late Pleistocene from Eastern Washington. In *Morphological Change in Quaternary Mammals of North America*, edited by Robert A. Martin and Anthony D. Barnosky, pp. 299–342. Cambridge: Cambridge University Press.

Rensberger, John M., Anthony D. Barnosky, and Patrick Spencer. 1984. *Geology and Paleontology of a Pleistocene-to-Holocene Loess Succession, Benton County, Washington.* Reports in Archaeology and History 100-39. Cheney: Eastern Washington University.

Richardson, Elmo R. 1968. Olympic National Park: 20 Years of Controversy. *Forest History* 12(1):6–15.

Rideout, Chester B. 1978. Mountain Goat. In *Big Game of North America: Ecology and Management*, edited by John L. Schmidt and Douglas L. Gilbert, pp. 149–59. Harrisburg, Penn.: Stackpole Books.

Rideout, Chester B., and Robert S. Hoffmann. 1975. *Oreamnos americanus. Mammalian Species* 63:1–6.

Robisch, Elena, and R. Gerald Wright. 1995. A Survey of Ungulate Management in Selected U.S. National Parks. *Natural Areas Jounral* 15:117–23.

Rochefort, Regina M., Ronda L. Little, Andrea Woodward, and David L. Peterson. 1994. Changes in Sub-alpine Tree Distribution in Western North America: A Review of Climatic and Other Causal Factors. *Holocene* 4:89–100.

Rochefort, Regina M., and David L. Peterson. 1996. Temporal and Spatial Distribution of Trees in Subalpine Meadows of Mount Rainier National Park, Washington, U.S.A. *Arctic and Alpine Research* 28:52–59.

Roloff, Clifford Edwin. 1934. The Mount Olympus National Monument. *Washington Historical Quarterly* 25(3):214–28.

Romesburg, H. Charles. 1981. Wildlife Science: Gaining Reliable Knowledge. *Journal of Wildlife Management* 45:293–313.

——. 1989. More on Gaining Reliable Knowledge: A Reply. *Journal of Wildlife Management* 53:1177–80.

Rust, Henry Judson. 1946. Mammals of Northern Idaho. *Journal of Mammalogy* 27:308–27.

Samson, Fred B., Francisco Perez-Trejo, Hal Salwasser, Leonard F. Ruggiero, and Mark L. Shaffer. 1985. On Determining and Managing Minimum Population Size. *Wildlife Society Bulletin* 13:425–33.

Savage, Howard. 1973. Faunal Changes through Time in British Columbia Coastal Sites and the Implications Thereof. Paper presented at the Canadian Archaeological Association Meeting, March 16, Burnaby, British Columbia.

Schalk, Randall F. 1988. The Evolution and Diversification of Native Land Use Systems on the Olympic Peninsula: A Research Design. Unpublished report on file, Olympic National Park, Port Angeles.

——. 1993. A Review of the Ethnographic and Archaeological Evidence Relating to Mountain Goats in the Olympic Mountains. Unpublished report on file, Olympic National Park, Port Angeles.

Schamberger, Melvin L. 1970. Mammals of Mount Rainier National Park. Ph.D. dissertation, Oregon State University, Corvallis.

Scheffer, Victor B. 1946. Mammals of the Olympic Peninsula, Washington. Unpublished report on file, Olympic National Park, Port Angeles.

——. 1974. *A Voice for Wildlife.* New York: Charles Scribner's Sons.

——. 1976. The Future of Wildlife Management. *Wildlife Society Bulletin* 4:51–54.

——. 1991. *The Shaping of Environmentalism in America.* Seattle: University of Washington Press.

——. 1993a. The Olympic Mountain Goat Controversy: A Perspective. *Conservation Biology* 7:916–19.

——. 1993b. Reply to the Anunsens. *Conservation Biology* 7:958.

——. 1995. *Mammals of the Olympic National Park and Vicinity.* Northwest Fauna No. 2. Olympia, Wash.: Society for Northwest Vertebrate Biology.

Scheffer, Victor B., and Walter W. Dalquest. 1944. Records of Mountain Goat and Moose from Washington State. *Journal of Mammalogy* 25:412–13.

Schreiner, Edward G. 1974. Vegetation Dynamics and Human Trampling in Three Subalpine Communities of Olympic National Park, Washington. M.S. thesis, University of Washington, Seattle.

——. 1982. The Role of Exotic Species in Plant Succession Following Human Disturbance in an Alpine Area of Olympic National Park.

Ph.D. dissertation, University of Washington, Seattle.

——. 1987. Vegetation in Relation to Non-native Mountain Goats in the Olympic Mountains. Unpublished report on file, Olympic National Park, Port Angeles.

——. 1994. Mountain Goat Herbivory. In *Mountain Goats in Olympic National Park: Biology and Management of an Introduced Species*, by Douglas B. Houston, Edward G. Schreiner, and Bruce B. Moorhead, pp. 105–88. Denver: USDI–NPS Scientific Monograph NPS/NROLYM/NRSM-94/25.

Schreiner, Edward G., Kirsten A. Kruger, Patricia J. Happe, and Douglas B. Houston. 1996. Understory Patch Dynamics and Ungulate Herbivory in Old-Growth Forests of Olympic National Park, Washington. *Canadian Journal of Forestry Research* 26:255–65.

Schreiner, E. G., and B. B. Moorhead. 1981. Human Impact Inventory and Backcountry Rehabilitation in Olympic NP: Research and Its Application. *Park Science* 1:1–4.

Schreiner, E. G., A. Woodward, and M. Gracz. 1993. Vegetation in Relation to Introduced Mountain Goats in Olympic National Park: A Technical Report. Unpublished report on file, Olympic National Park, Port Angeles.

Schullery, Paul. 1984. A History of Native Elk in Mount Rainier National Park. Unpublished report on file, Mount Rainier National Park, Ashford, Washington.

Schulting, Rick. 1994. The Hair of the Dog: The Identification of a Coast Salish Dog-Hair Blanket from Yale, British Columbia. *Canadian Journal of Archaeology* 18:57–76.

Schultz, Susan. 1993. A Review of the Historical Evidence Relating to Mountain Goats in the Olympic Mountains Prior to 1925. Unpublished draft report on file, Olympic National Park, Port Angeles.

——. 1994. Review of the Historical Evidence Relating to Mountain Goats in the Olympic Mountains before 1925. In *Mountain Goats in Olympic National Park: Biology and Management of an Introduced Species*, by Douglas B. Houston, Edward G. Schreiner, and Bruce B. Moorhead, pp. 256–93. Denver: USDI–NPS Scientific Monograph NPS/NROLYM/NRSM-94/25.

Sellars, Richard W. 1989. Science or Scenery? A Conflict of Values in National Parks. *Wilderness* 52(185):28–39.

——. 1992. The Roots of National Park Management. *Journal of Forestry* 90(1):16–19.

——. 1993. The Rise and Decline of Ecological Attitudes in National Park Management, 1929–1940, Part III (Conclusion): Growth and Diversification of the National Park Service. *George Wright Forum* 10(3):38–54.

Seton, E. T. 1909. *Lives of Game Animals, Vol. 3: Hoofed Mammals.* New York: Literary Guild of America.

Shrader-Frechette, Kristin S., and Earl D. McCoy. 1995. Natural Landscapes, Natural Communities, and Natural Ecosystems. *Forest and Conservation History* 39:138–42.

Sinclair, A. R. E. 1991. Science and the Practice of Wildlife Management. *Journal of Wildlife Management* 55:767–73.

Sinclair, William J. 1903. A Preliminary Account of the Exploration of Potter Creek Cave, Shasta County, California. *Science* 17:708–12.

——. 1904. The Exploration of the Potter Creek Cave. *University of California Publications in American Archaeology and Ethnology* 2:1–27.

——. 1905. New Mammalia from the Quaternary Caves of California. *University of California Department of Geology Bulletin* 4:145–61.

Sleeper, Barbara L. 1990. Out on a Ledge. *Animals* 23(4):18–25.

Sleeper, Barbara L., and Michael Hutchins. 1989. The Horns of a Dilemma. *Pacific Discovery* 42(1):16–24.

Smith, Bruce L. 1977. Influence of Snow Conditions on Winter Distribution, Habitat Use, and Group Size of Mountain Goats. In *Proceedings of the First International Mountain Goat Symposium,* edited by William Samuel and W. G. Macgregor, pp. 174–189. Victoria, B.C.: Province of British Columbia, Ministry of Recreation and Conservation, Fish and Wildlife Branch.

Smith, Christian A., and Kenneth J. Raedeke. 1982. Group Size an Movements of a Dispersed, Low Density Goat Population with Comments on Inbreeding and Human Impact. *Proceedings of the Biennial Symposium of the Northern Wild Sheep and Goat Council* 3:54–67.

Smith, Lorraine C. 1976. Communication via Scientific Writing. *Canadian Field-Naturalist* 90:101–2.

Smith, T. B., M. W. Bruford, and R. K. Wayne. 1993. The Preservation of Process: The Missing Element in Conservation. *Biodiversity Letters* 1:164–67.

Soulé, Michael E. 1987a. Introduction. In *Viable Populations for Conservation,* edited by Michael E. Soulé, pp. 1–10. Cambridge: Cambridge University Press.

—— (ed.). 1987b. *Viable Populations for Conservation.* Cambridge: Cambridge University Press.

Steadman, David W. 1995. Prehistoric Extinctions of Pacific Island Birds: Biodiversity Meets Zooarchaeology. *Science* 267:1123–31.

Stenseth, Nils C., and J. Maynard Smith. 1984. Coevolution in Ecosystems: Red Queen Evolution or Stasis? *Evolution* 38:870–80.

Sterling, Keir B. 1989. Builders of the U.S. Biological Survey, 1885–1930. *Journal of Forest History* 33(4):180–87.

Stevens, Victoria. 1980. Final Report, Terrestrial Baseline Surveys, Nonnative Mountain Goats in the Olympic National Park. Unpublished report on file, Olympic National Park, Port Angeles.

——. 1982. Factors Reflecting Mountain Goat Condition and Habitat Quality: A Comparison of Sub-populations in Olympic National Park. In *Ecological Research in National Parks of the Pacific Northwest,* edited by Edward E. Starkey, Jerry F. Franklin, and Jean W. Matthews, pp. 51–57. Corvallis: Oregon State University Forest Research Laboratory.

——. 1983. The Dynamics of Dispersal in an Introduced Mountain Goat Population. Ph.D. dissertation, University of Washington, Seattle.

Stevens, Victoria, and Charles Driver. 1978. Initial Observations on a Tagged Mountain Goat Population in the Olympic Mountains. *Proceedings of the Biennial Symposium of the Northern Wild Sheep and Goat Council* 1:165–73.

Stewart, Frances L., and Kathlyn M. Stewart. 1996. The Boardwalk and Grassy Bay Sites: Patterns of Seasonality and Subsistence on the Northern Northwest Coast, B. C. *Canadian Journal of Archaeology* 20:39–60.

Stock, Chester. 1936. A New Mountain Goat from the Quaternary of Smith Creek Cave, Nevada. *Bulletin of the Southern California Academy of Sciences* 35:149–53.

Stohlgren, Thomas J., and James F. Quinn. 1992. An Assessment of Biotic Inventories in Western U.S. National Parks. *Natural Areas Journal* 12:145–54.

Stone, Charles P., and Lloyd L. Loope. 1996. Alien Species in Hawaiian National Parks. In *Science and Ecosystem Management in the National Parks,* edited by William L. Halvorson and Gary E. Davis, pp. 132–58. Tucson: University of Arizona Press.

Stottlemyer, J. Robert. 1981. Evolution of Management Policy and Research in the National Parks. *Journal of Forestry* 79:16–20.

Suckley, George, and George Gibbs. 1860. Report upon the Mammals Collected on the Survey, Chapter III. *Reports of Explorations and and Surveys, to Ascertain the Most Practicable and Economical Route for a Railroad from the Mississippi River to the Pacific Ocean,* vol. 12, book 2, pp. 107–39. Washington, D.C.: Government Printing Office.

Suttles, Wayne. 1990a. Central Coast Salish. In *Handbook of North American Indians, Vol. 7: Northwest Coast,* edited by Wayne Suttles, pp. 453–75. Washington, D.C.: Smithsonian Institution Press.

—— (ed.). 1990b. *Handbook of North American Indians, Vol. 7: Northwest Coast.* Washington, D.C.: Smithsonian Institution Press.

Suttles, Wayne, and Barbara Lane. 1990. Southern Coast Salish. In *Handbook of North American Indians, Vol. 7: Northwest Coast,* edited by Wayne Suttles, pp. 485–502. Washington, D.C.: Smithsonian Institution Press.

Svihla, Arthur, and Ruth D. Svihla. 1933. Mammals of Clallam County, Washington. *Murrelet* 14:37–41.

Swarth, Harry S. 1936. Origins of the Fauna of the Sitkan District, Alaska. *Proceedings of the California Academy of Sciences,* fourth series 23(3):59–78.

Sweeney, James M., and Peter W. Stangel. 1995. Walking the Line: Science versus Advocacy. *Transactions of the Sixtieth North American Wildlife and Natural Resources Conference,* pp. 216–20. Washington, D.C.: Wildlife Management Institute.

Taulman, James F. 1989. Hoary Marmots Attracted to Human Urine Sites. *Northwestern Naturalist* 70:36–37.

Taylor, Walter P., and William T. Shaw. 1927. *Mammals and Birds of Mount Rainier National Park.* Washington, D.C.: U.S. Government Printing Office.

——. 1929. *Provisional List of Land Mammals of the State of Washington.* Occasional Papers of the Charles R. Conner Museum No. 2. Pullman: State College of Washington (Washington State University).

Thornton, Russell. 1997. Aboriginal North American Population and Rates of Decline, ca. A.D. 1500–1900. *Current Anthropology* 38:310–15.

Tomich, P. Quentin. 1982. Ground Squirrels: *Spermophilus beecheyi* and Allies. In *Wild Mammals of North America: Biology, Management, Economics,* edited by Joseph A. Chapman and George A. Feldhamer, pp. 192–208. Baltimore: Johns Hopkins University Press.

Tracy, C. Richard, and T. Luke George. 1992. On the Determinants of Extinction. *American Naturalist* 139:102–22.

Trombulak, Stephen C. 1988. *Spermophilus saturatus. Mammalian Species* 322:1–4.

Truett, Joe. 1996. Bison and Elk in the American Southwest: In Search of the Pristine. *Environmental Management* 20:195–206.

Tuler, Seth, and Chuck Janda. 1991. Olympic NP Mountain Goat Removal Project Subject of Risk Assessment Report. *Park Science* 11(1):14–15.

Tuler, Seth, Gary E. Machlis, and Roger E. Kasperson. 1992. Mountain Goat Removal in Olympic National Park: A Case Study of the Role of Organizational Culture in Individual Risk Decisions and Behavior. *RISK—Issues in Health and Safety* 3:317–40.

Twight, Ben W. 1983. *Organizational Values and Political Power: The Forest Service versus the Olympic National Park.* University Park: Pennsylvania State University Press.

U.S. Department of the Interior–National Park Service. 1970. *Management Policies.* Washington, D.C.: U.S. Department of the Interior.

——. 1978. *Management Policies.* Washington, D.C.: U.S. Department of the Interior.

——. 1988. *Management Policies.* Washington, D.C.: U.S. Department of the Interior.

van Zyll de Jong, C. G. 1983. *Handbook of Canadian Mammals, 1: Marsupials and Insectivores.* Ottawa: National Museums of Canada.

Vaughan, Michael R. 1975. Aspects of Mountain Goat Ecology, Wallowa Mountains, Oregon. M.S. thesis, Oregon State University, Corvallis.

Vrba, Elisabeth S. 1992. Mammals as a Key to Evolutionary Theory. *Journal of Mammalogy* 73:1–28.

Wagner, Frederic H. 1989. American Wildlife Management at the Crossroads. *Wildlife Society Bulletin* 17:354–60.

——. 1996. Ethnics, Science, and Public Policy. *BioScience* 46:765–86.

Wagner, Frederic H., Ronald Foresta, R. Bruce Gill, Dale R. McCullough, Michael R. Pelton, William F. Porter, and Hal Salwasser. 1995. *Wildlife Policies in the U.S. National Parks.* Washington, D.C.: Island Press.

Waitt, Richard B., Jr., and Robert M. Thorson. 1983. The Cordilleran Ice Sheet in Washington, Idaho, and Montana. In *Late Quaternary Environments of the United States, Vol. 1, The Late Pleistocene,* edited by

Stephen C. Porter, pp. 53–70. Minneapolis: University of Minnesota Press.

Walters, C. 1986. *Adaptive Management of Renewable Resources.* New York: Macmillan.

Warner, Barry G., Rolf W. Mathewes, and John J. Clague. 1982. Ice-Free Conditions on the Queen Charlotte Islands, British Columbia, at the Height of Late Wisconsin Glaciation. *Science* 218:675–77.

Warren, Hank [Henry C.]. 1987. Unpublished notes from Ed Schreiner's [March] goat briefing for park staff on file, Olympic National Park, Port Angeles.

Warren, Henry C. 1982. *Olympic: The Story behind the Scenery.* Las Vegas: KC Publications.

Warren, R. J., R. A. Fayrer-Hosken, D. A. Jessup, R. A. Garrott, and J. F. Kirkpatrick. 1992. The Applicability of Contraceptives in the Elimination or Control of Exotic Mountain Goats from Olympic National Park. Unpublished report on file, Olympic National Park, Port Angeles.

Washington Writers' Program. 1950. *The New Washington: A Guide to the Evergreen State.* Revised edition. Portland: Binfords and Mort.

Wauer, Roland H., and William R. Supernaugh. 1983. Wildlife Management in the National Parks. *National Parks* 57(7–8):12–16.

Webster, E. B. 1920. *The King of the Olympics: The Roosevelt Elk and Other Mammals of the Olympic Mountains.* Port Angeles: Privately published.

——. 1932. Status of Mountain Goats Introduced into the Olympic Mountains, Washington. *Murrelet* 13:25.

Webster, R., and D. H. Yaalon. 1994. The Research Paper: An Informal Guide for Authors. *Catena* 21:3–11.

Weingrod, Carmi. 1994. On the Horns of a Dilemma. *National Parks* 68(5–6):29–32.

Weisbrod, A. R. 1979. Insularity and Mammal Species Number in Two National Parks. In *Proceedings of the First Conference on Scientific Research in the National Parks,* vol. 2, edited by Robert M. Linn, pp. 83–87. Transactions and Proceedings Series No. 5. Washington, D.C.: National Park Service.

Weisbrod, A. R., and J. A. Dragavon. 1979. Distribution and Richness of Mammal Species in Mount Rainier National Park. In *Proceedings of the First Conference on Scientific Research in the National Parks,* vol. 2, edited by Robert M. Linn, pp. 315–18. Transactions and Proceedings Series No. 5. Washington, D.C.: National Park Service.

Whitlock, Cathy. 1992. Vegetational and Climatic History of the Pacific Northwest during the Last 20,000 Years: Implications for Understanding Present-day Biodiversity. *Northwest Environmental Journal* 8:5–28.

Wickersham, James. 1961. A National Park in the Olympics ... 1890. *Living Wilderness* 77:5–13.

Wigal, Ronald A., and Victor L. Coggins. 1982. Mountain Goat (*Oreamnos americanus*). In *Wild Mammals of North America: Biology, Management, Economics,* edited by Joseph A. Chapman and George A. Feldhamer, pp. 1008–20. Baltimore: Johns Hopkins University Press.

Wilson, Cynthia, Richard E. Johnson, and James D. Reichel. 1980. New Records for the Northern Bog Lemming in Washington. *Murrelet* 61:104–6.

Wister, Owen. 1904. The White Goat and His Ways. In *Musk-Ox, Bison, Sheep and Goat,* edited by Caspar Whitney, pp. 227–75. New York: Macmillan.

Wood, Robert L. 1967. *Across the Olympic Mountains: The Press Expedition, 1889–1890.* Seattle: University of Washington Press.

——. 1968. *Trail Country: Olympic National Park.* Seattle: The Mountaineers.

Woods, S. E., Jr. 1980. *The Squirrels of Canada.* Ottawa: National Museums of Canada.

Woodward, A., E. G. Schreiner, D. B. Houston, and B. B. Moorhead. 1994. Ungulate–Forest Relationships in Olympic National Park: Retrospective Exclosure Studies. *Northwest Science* 68:97–110.

Woodward, James, and David Goodstein. 1996. Conduct, Misconduct and the Structure of Science. *American Scientist* 84:479–90.

Wright, George M., Joseph B. Dixon, and Ben H. Thompson. 1933. *Fauna of the National Parks of the United States: A Preliminary Survey of Faunal Relations in National Parks.* Contribution of Wild Life Survey, Fauna Series No. 1. Washington, D.C.: U.S. Government Printing Office.

Wright, George M., and Ben H. Thompson. 1935. *Fauna of the National Parks of the United States: Wildlife Management in the National Parks.* Contribution of Wild Life Survey, Fauna Series No. 2. Washington, D.C.: U.S. Government Printing Office.

Wright, Karen. 1996. Getting the Goats. *Scientific American* 274(3): 22–23, 26.

Wright, R. Gerald. 1992. *Wildlife Research and Management in the National Parks.* Urbana: University of Illinois Press.

Wuerthner, George. 1996. Alien Invasion. *National Parks* 70(11–12): 32–35.

Wunderlich, Robert C., Brian D. Winter, and John H. Meyer. 1994. Restoration of the Elwha River Ecosystem. *Fisheries* 19(8):11–19.

Zube, Ervin H. 1996. Management in National Parks: From Scenery to Science. In *Science and Ecosystem Management in the National Parks,* edited by William L. Halvorson and Gary E. Davis, pp. 11–22. Tucson: University of Arizona Press.

INDEX